Advanced Textbooks in Control and Signal Processing

T0142105

Springer
London
Berlin
Heidelberg
New York
Hong Kong
Milan
Paris
Tokyo

Series Editors

Professor Michael J. Grimble, Professor of Industrial Systems and Director
Professor Michael A. Johnson, Professor of Control Systems and Deputy Director

Industrial Control Centre, Department of Electronic and Electrical Engineering,
University of Strathclyde, Graham Hills Building, 50 George Street, Glasgow G1 1QE, U.K.

Other titles published in this series:

M. Nørgaard, O. Ravn, N.K. Poulsen
and L.K. Hansen

Neural Networks for Modelling and Control of Dynamic Systems

A Practitioner's Handbook

With 84 Figures

Springer

M. Nørgaard, MSc, EE, PhD
Department of Automation, Technical University of Denmark, Building 326,
DK-2800, Lyngby, Denmark

O. Ravn, MSc, EE, PhD
Department of Automation, Technical University of Denmark, Building 326,
DK-2800, Lyngby, Denmark

N.K. Poulsen, MSc, EE, PhD
Department of Mathematical Modelling, Technical University of Denmark,
Building 321, DK-2800, Lyngby, Denmark

L.K. Hansen, MSc, PhD
Department of Mathematical Modelling, Technical University of Denmark,
Building 321, DK-2800, Lyngby, Denmark

ISBN 1-85233-227-1 Springer-Verlag London Berlin Heidelberg

British Library Cataloguing in Publication Data
Neural networks for modelling and control of dynamic
   systems : a practitioner's handbook. - (Advanced textbooks
   in control and signal processing)
   1.Neural networks (Computer science) 2.Automatic control
   I.Norgaard, M.
   629.8'9
   ISBN 1852332271

Library of Congress Cataloging-in-Publication Data
Neural networks for modelling and control of dynamic systems : a practitioner's
handbook / M. Nørgaard ... [et al.].
      p. cm. -- (Advanced textbooks in control and signal processing)
   Includes bibliographical references and index.
   ISBN 1-85233-227-1 (alk. paper)
   1. Neural networks (Computer science) 2. Computer simulation. 3. Automatic control.
   4.Nonlinear theories. I. Nørgaard, Magnus. II. Series.
   QA76.87 N4847 2000
   006.3'2--dc21   99-049801

© Springer-Verlag London Limited 2000
Printed in Great Britain
2nd printing 2001
3rd printing, with corrections 2003

MATLAB® is the registered trademark of The MathWorks, Inc., http://www.mathworks.com

Typesetting: Camera ready by authors
Printed and bound at the Athenæum Press Ltd., Gateshead, Tyne and Wear
69/3830-5432 Printed on acid-free paper SPIN 10911488

Series Editors' Foreword

The topics of control engineering and signal processing continue to flourish and develop. In common with general scientific investigation, new ideas, concepts and interpretations emerge quite spontaneously and these are then discussed, used, discarded or subsumed into the prevailing subject paradigm. Sometimes these innovative concepts coalesce into a new sub-discipline within the broad subject tapestry of control and signal processing. This preliminary battle between old and new usually takes place at conferences, through the Internet and in the journals of the discipline. After a little more maturity has been acquired by the new concepts then archival publication as a scientific or engineering monograph may occur.

A new concept in control and signal processing is known to have arrived when sufficient material has developed for the topic to be taught as a specialised tutorial workshop or as a course to undergraduates, graduates or industrial engineers. The *Advanced Textbooks in Control and Signal Processing* series is designed as a vehicle for the systematic presentation of course material for both popular and innovative topics in the discipline. It is hoped that prospective authors will welcome the opportunity to publish a structured presentation of either existing subject areas or some of the newer emerging control and signal processing technologies.

This is a fascinating and well-written book. According to the authors: *Neural Networks constitutes a very large research field, and it is difficult to obtain a clear overview of the entire field*. This book threads a careful way through that field to guide the reader to the items necessary to use neural networks in system identification and ultimately in control systems applications. So if you wanted to know *Why use Neural Networks?* then this is the book for you.

The book comes with plenty of added features. There is neural network software available at an associated website and an e-mail address to further the dialogue on the contents of the book itself. There is even a challenge, for early in the book the authors state their belief that it is possible to reduce development time and achieve better performance using neural networks as opposed to using auto-tuned PID for general purpose controllers. It might be interesting for some enterprising student to put that belief to the test and assess the advantages and disadvantages of both approaches on some benchmark problems.

In conclusion a very welcome addition to the *Advanced Textbooks in Control and Signal Processing* series and a nice complementary book to another in the series, that of Kim Man, and his colleague on *Genetic Algorithms* (ISBN 1-85233-072-4).

M.J. Grimble and M.A. Johnson
Industrial Control Centre
Glasgow, Scotland, U.K.
November 1999

Preface

Aim of the book. The main goal of this book is to describe approaches to neural-network-based control that are found to be practically applicable to a reasonably wide class of unknown nonlinear systems. System identification is an integral part of such a control system design and consequently it calls for considerable attention as well. The system identification is necessary to establish a model based on which the controller can be designed, and it is useful for tuning and simulation before applying the controller to the real system. However, system identification is relevant in many other applications; e.g., simulation, prediction, and fault detection. For this reason the theoretical explorations in the book have been split into two main sections that are weighted approximately equally: a section about system identification and a section about design of control systems. Although the treatment of system identification has a certain bias towards the application to control, it has been written so that expert knowledge about control theory is not necessary.

In writing the book an attempt has been made to outline a feasible path through the "jungle" of neural network solutions. The emphasis is on guidelines for working solutions and on practical advice on implementation issues. A completely automatic procedure for system identification and control system design is not realistic. Thus, an attempt has been made to provide techniques that minimize the effort required by the user and leave it up to him/her to answer only a few reasonably well-defined assessment questions. Algorithms are detailed to be fast and numerically sound, but the book does not go as far as to describe the actual programming. The necessary theoretical background is given for the methods presented, but the reader will not find rigorous mathematical proofs for the statements.

Approach. The philosophy underlying the selection of topics for the book is that a pragmatic approach is the road to success. It is believed that one of the most important lessons to be learned from the numerous automatic control applications developed over the past half century is that simple solutions actually solve most problems quite well. Regardless of the fact that all systems to some extent exhibit a nonlinear behavior, it turns out that they can often be controlled satisfactorily with simple linear controllers. When neural networks are introduced as a tool for improving the performance of control systems for a general class of unknown nonlinear systems, it should be done in the same spirit. Thus, most of the theory is derived in a somewhat heuristic fashion. It has been important to pursue methods that yield good performance in practice and thus there is generally little concern with the possibility for proving stability. In writing the book we have avoided becoming too deeply absorbed in the mathematical and statistical foundation for the neural-network-based methods. The reader will find many mathematical derivations, but we have tried not to bring in more theory than is needed to provide the reader with the insight to understand the principles behind the methods and the detail enabling an implementation.

A consequence of this philosophy is that the focus is placed on two-layer perceptron neural networks with hyperbolic tangent hidden units and linear output units. This is probably the most commonly used network architecture as it works quite well in many practical applications. While the implementation details are adapted to these networks, most theoretical explorations apply directly to neural networks in general. However, the reader is referred to more fundamental textbooks on neural networks for a treatment of other types of neural networks.

Supporting software. The book provides the theoretical foundation for two sets of tools for the mathematical software package MATLAB®:

- The NNSYSID Toolbox, which contains a collection of MATLAB® functions for system identification with neural networks.

- The NNCTRL Toolkit, which contains a set of tools for design and simulation of neural-network-based controllers.

Detailed description of the contents and use of the two packages can be found in the manuals Nørgaard (1997) and Nørgaard (1996a). Software and manuals can be downloaded from the internet at:
http://www.iau.dtu.dk/nnspringer.html

On this Web page there is additional supporting material for the book as well. For example, assignments, simulation models, list of errors, etc. The authors would like to encourage the readers to provide additional material for the web page or to give feedback on the book by using the e-mail address:
nnspringer@iau.dtu.dk

Prerequisites. Primarily, this book addresses engineering students at the graduate level. It is appropriate as a textbook in a course that mixes theory with practical laboratory sessions. It is also useful as a handbook in practical projects and courses. Engineering professionals should find the book relevant as well; either for self-study or as a handbook for implementation.

An introductory course in adaptive control is considered the most appropriate prerequisite for the book. As a minimum the reader should know about matrix calculus, basic statistics, system identification/time series analysis, and digital control (Chapter 3). It is an advantage if the reader is acquainted with the neural network field, but it is not a vital prerequisite for understanding the material.

Outline of the book. The book has four chapters:

Chapter 1 introduces the multilayer perceptron neural network and discusses why and when it is relevant to use it in system identification and control system design.

Chapter 2 outlines a procedure for system identification with neural networks and proceeds with a thorough treatment of each stage in the procedure. The covered issues encompass:

- Experiment design. How to conduct an experiment to collect a set of data for neural network modelling.

- Model structure selection. Conventional linear model structures are extended to nonlinear systems by incorporation of neural networks. Automatic methods for selection of neural network architectures, so-called pruning algorithms, are also described.

- Neural network training. It is described how to train networks as models of dynamic system with a prediction error method. Optimization methods relevant for neural network training are described. Regularization by weight decay is introduced as an extension of the basic prediction error method.

- Validation. Techniques for assessing neural network models are treated and concepts like generalization error and average generalization error are introduced

- The chapter is concluded by outlining a set of rules of thumb for system identification with neural networks.

Chapter 3 provides an overview of a wide range of approaches to neural-network-based control system design. The features of each particular design are explored and the implementation issues treated. Some designs are characterized by being restricted to a relatively limited class of systems but simple to implement. Others are characterized by being applicable to a wider class of systems, but more difficult to implement. The designs have been divided into two categories:

- Direct design. The controller is in itself a neural network. Examples of designs in this category are direct inverse control, internal model control, feedback linearization, feedforward, and optimal control.

- Indirect design. Designs based on a neural network model of the system to be controlled. The controller is in this case either a time-varying linear controller or an optimization algorithm. Two approaches to indirect design are described. One is to use a conventional design based on models obtained by linearization of the neural network model. The second is a nonlinear predictive control design based on on-line optimization.

A benchmark system is used for illustrating the properties of the designs. The chapter is concluded by providing some guidelines for selecting the most appropriate controller for a particular application.

Chapter 4 describes four case studies in the use of neural networks for system identification and control. Each case illustrates several topics covered in the book.

Background of the work. The main portion of the material used in this book comes from the Ph.D. thesis written by the first author (Nørgaard, 1996b). The work was carried out at the Department of Automation, Technical University of Denmark with the other three authors as supervisors. In the Ph.D.-study the previously mentioned software tools were developed to demonstrate that practical use of the described methods was in fact possible. The tools were made available on the internet in 1994 and 1995, respectively, and since then thousands have downloaded and used the software. Many users have been eager to understand the underlying theory and have requested the thesis. The overwhelming interest in the work made the authors approach Springer-Verlag, London in order to make the material available to an even wider audience. Compared with the Ph.D. thesis, the text has been revised, some issues are no longer covered, and a few new sections have been added.

Acknowledgments. The authors would like to thank several people who have helped in the creation of the book: Kevin Wheeler for proofreading the manuscript, Paul Haase Sørensen for providing the material for the pneumatic servomechanism in Section 4.3, Svante Gunnarsson for providing material and data for the hydraulic crane example in Section 4.2, and Egill Rostrup for providing the fMRI data presented in Chapter 1.

Lyngby   *Magnus Nørgaard*

November 1999   *Ole Ravn*

Niels Kjølstad Poulsen

Lars Kai Hansen

Contents

1. Introduction

Many of the abilities one possesses as a human have been learned from examples. Thus, it is only natural to try to carry this "didactic principle" over to a computer program to make it learn how to output the desired answer for a given input. In a sense the artificial neural network is one such computer program; it is a mathematical formula with several adjustable parameters, which are tuned from a set of examples. These examples represent what the network should output when it is shown a particular input.

The book deals with two specific neural network applications: modelling of dynamic systems and control. Despite the fact that "learning from examples" sounds easy, many have been surprised to experience that often it is in fact extremely difficult to obtain working neural network solutions. It is hoped that the methods and recommendations given in this book will guide the reader to many successful neural network implementations.

1.1 Background

Today automatic control systems have become an integrated part of our everyday life. They appear in everything from simple electronic household products to airplanes and spacecrafts (see Figure 1.1). Automatic control systems can take highly different shapes but common to them all is their function to manipulate a system so that it behaves in a desired fashion. When designing a controller for a particular system, it is obvious that a vital intermediate step is to acquire some knowledge about how the system will respond when it is manipulated in various ways. Not until such knowledge is available, can one plan how the system should be controlled to exhibit a certain behavior.

1.1.1 Inferring Models and Controllers from Data

A common and practically oriented approach to control system design is to use physical insights about the system supplemented with a series of practical closed-loop tests. In the tests different design parameters are tried until a working controller is obtained. Another often-used approach is based on conducting a simple experiment with the system to provoke a particular response. Based on the knowledge of how this particular response is obtained, simple rules of thumb subsequently explain how the automatic control system should be designed (Ziegler and Nichols, 1942). Such procedures have even been automated in commercially available devices known as *auto-tuners*. Sometimes,

Figure 1.1. The NF-15B research aircraft. Boeing and the NeuroEngineering Group at NASA Ames Research Center have developed an adaptive neural-network-based flight control system capable of recovering from damages due to failures or serious accidents. The control system has been tested on the NF-15B at NASA Dryden Flight Research Center. The F-15 has been modified so that damages can be simulated by activating the canards (the small front wings), thereby changing the airflow over the main wings. (NASA Photo by Tony Landis, Dryden Flight Research Center.)

simple design approaches such as those just cited, are not adequate, either because they simply fail to work or because demands in performance are too strong to be satisfied by means of simple rules of thumb. In such cases more advanced design methods must be considered. These designs will in general require that knowledge about the system to be controlled is more structured in that it should be specified in terms of differential or difference equations. A mathematical description of this kind is called *a model* of the system. Basically, there are two ways in which a model can be established: it can be derived in a deductive manner using laws of nature, or it can be inferred from a set of data collected during a practical experiment with the system.

The first method can be simple, but in most cases it is excessively time-consuming. This is possibly the most time-consuming stage in the controller design. Not infrequently it may even be considered unrealistic or impossible to obtain a sufficiently accurate model in this way. The second method, which is commonly referred to as *system identification*, in these situations can be a useful short cut for deriving mathematical models. Although system identification not always results in equally accurate models, a satisfactory model can often be obtained with a reasonable effort. The main drawback is the requirement to conduct a practical experiment that brings the system through its entire range of operation. Also, a certain knowledge about the system is still required.

System identification techniques are widely used in relation to control system design and many successful applications have been made over the years. Sometimes system identification is even implemented as an integral part of the controller. This is known as an *adaptive controller* and it is typically designed to control systems whose dynamical characteristics vary with time. In the typical adaptive controller a model that is valid under the current operating conditions is identified on-the-fly, and the controller is then redesigned in agreement with the current model.

Much literature is available on system identification, adaptive control, and control system design in general, but traditionally most of it has focused on dealing with models and controllers described by linear differential or difference equations. However, motivated by the fact that all systems exhibit some kind of nonlinear behavior, there has recently been much focus on different approaches to nonlinear system identification and controller design. One of the key players in this endeavor is the *artificial neural network*. Artificial neural networks represent a discipline that originates from a desire to imitate the functions of a biological neural network, namely the brain. Artificial neural networks, or just *neural networks*, as they are most often abbreviated, have been one of the major buzz words in the recent years. Apart from system identification and control they have been applied in such diverse fields as insurance, medicine, banking, speech recognition, and image processing to mention just a very few examples. They are typically implemented in software, but dedicated neural network hardware is also available for increased execution speed.

Neural networks constitute a very large research field, and it is difficult to obtain a clear overview of the entire field. Several motives originally lead researchers to study neural networks. One of the primary motives was to create a computer program that was able to learn from experience. The hope was to create an alternative to conventional programming techniques, where rules were coded directly into the computer. When the experience mentioned is interpreted as knowledge about how certain inputs affect a system, it is obvious that neural networks must have something in common with the techniques applied in system identification and adaptive control.

As the research on neural networks has evolved, more and more types of networks have been introduced while still less emphasis is placed on the connection to the biological neural network. In fact, the neural networks that are most popular today have very little resemblance to the brain, and one might argue that it would be fairer to regard them simply as a discipline under statistics. These neural networks are vehicles that in a generic sense can learn nonlinear mappings from a set of observations. Neural networks are not the only technique available for approximating such generic nonlinear mappings; the list over similar techniques is in fact quite long. See Sjöberg et al. (1995) for some examples that are relevant to system identification.

1.1.2 Why Use Neural Networks?

Why have neural networks attracted particular attention compared with alternative techniques? For a given application it is of course difficult to say that one identification technique will outperform another before they have both been evaluated. Nevertheless, it is desirable to consider only one technique for all applications rather than having to evaluate several candidates on each new application. Partly because it simplifies the modelling process itself, and also because it will enable implementation of generic tools for control system design. When searching for a single technique that in most cases of practical interest performs reasonably well, certain types of neural network appear to be an excellent choice. In particular the multilayer perceptron network has gained an immense popularity. From numerous practical applications published over the past decade there seems to be substantial evidence that multilayer perceptrons indeed possess an impressive ability. Lately, there have also been some theoretical results that attempt to explain the reasons for this success. For supplementary information one may consult Barron (1993) and Juditsky et al. (1995).

How are neural networks useful for control system design? It is practical to distinguish between the following two categories of controllers:

Highly specialized controllers that are relevant when the system to be controlled is in some sense difficult to stabilize or when the performance is extremely important.

General purpose controllers where the same controller structure can be used on a wide class of practical systems. The controllers are characterized by being simple to tune so that a satisfactory performance can be achieved with a modest effort.

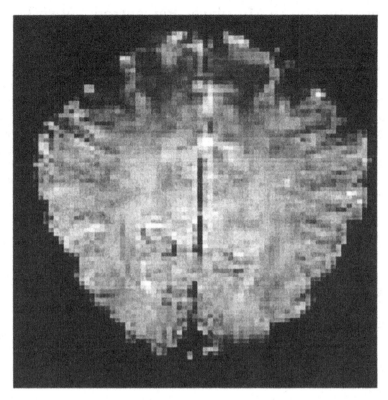

Figure 1.2. Functional magnetic resonance imaging is a new non-invasive window to the working human brain. The magnetic resonance signal is sensitive to minute changes in blood perfusion, hence, an indirect measure of the neuronal activity in the brain tissue. In neuroinformatics research, artificial neural networks are used for decoding the spatio-temporal distribution of activity that follows from external stimuli Mørch et al. (1997). The picture shows a slice through the primary visual areas in the lower back of the brain. Such slices are acquired at a rate of three images per second. (By courtesy of Dr. Egill Rostrup, Danish Research Centre for Magnetic Resonance, Hvidovre Hospital, Denmark.)

Basically, neural networks are relevant in both cases, but they probably have the biggest potential within general purpose control. It is believed that their ability to model a wide class of systems in many applications can reduce time spent on development and offer a better performance than can be obtained with conventional techniques like auto-tuned PID-controllers.

1.2 Introduction to Multilayer Perceptron Networks

When scanning the vast amount of literature on neural networks there does not seem to be one definition of a neural network with which everyone agrees, but most types of neural networks can be covered by the definition: *A system of simple processing elements, neurons, that are connected into a network by a set of (synaptic) weights.* The function of the network is determined by the architecture of the network, the magnitude of the weights and the processing element's mode of operation. It is often stressed that the processing elements should act in parallel as the neurons do in the brain, but when considering some of the most popular network types it is not quite clear exactly how the word "parallel" should in fact be interpreted.

1.2.1 The Neuron

The *neuron* or *node* or *unit*, as it is also called, is a processing element that takes a number of inputs, weights them, sums them up, and uses the result as the argument for a singular valued function, the *activation function*. See the illustration in Figure 1.3.

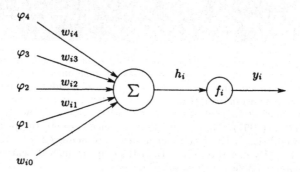

Figure 1.3. A neuron: $y_i = f_i(h_i) = f_i \left(\sum_{j=1}^{n} w_{i,j} + w_{i,0} \right)$.

The inputs to a unit can either be outputs of other units or they can be external inputs. The displacement $w_{i,0}$ is called the *bias* and can be interpreted as a weight applied to a pseudo input which is clamped to the constant value 1. Essentially, the activation function f_i can take any form but most often it is monotonic. Figure 1.4 displays some common activation functions.

In this book only the linear and the hyperbolic tangent activation functions (a and b) will be used. The sigmoid function (c) has historically been a very popular choice (Rumelhart and McClelland, 1986) but since it is related to

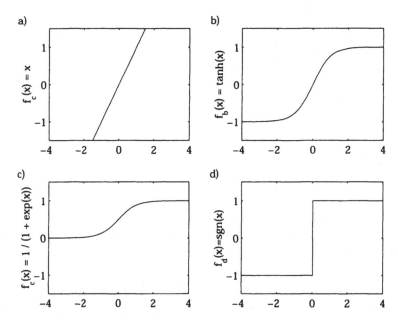

Figure 1.4. Four different activation functions. a) linear: $f_a(x) = x$; b) hyperbolic tangent: $f_b(x) = \tanh(x)$; c) sigmoid: $f_c(x) = 1/(1 + \exp(x))$; d) step: $f_d(x) = \text{sgn}(x)$.

the *tanh* by the simple transformation $F_c = (F_b + 1)/2$, it makes no difference which of these is used. The step function (d) is of particular importance to networks used for categorization, e.g., networks applied for providing yes/no answers. It is common to call a unit with a hyperbolic tangent (or linear or sigmoid) activation function *a hyperbolic (or linear or sigmoid) unit*.

1.2.2 The Multilayer Perceptron

Units can be combined into a network in numerous fashions. Several text-books are available that give give an introduction to various network archi-tectures (Hertz et al., 1991; Haykin, 1998; Zurada, 1992). Beyond any doubt, the most common of these is the *multilayer perceptron (MLP) network*. The basic MLP-network is constructed by ordering the units in layers, letting each unit in a layer take as input only the outputs of units in the previous layer or external inputs. If the network has two such layers of units, it is referred to as a two layer network, if it has three layers it is called a three layer network and so on. Due to the structure, this type of network is often referred to as a *feedforward network*. An example is depicted in Figure 1.5.

The second layer in Figure 1.5 is called the *output layer* referring to the fact that it produces the output of the network. The first layer is known as the

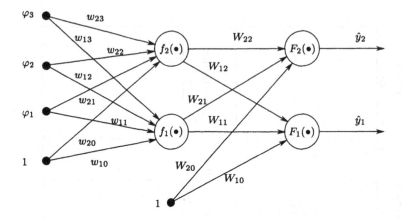

Figure 1.5. A fully connected two layer feedforward network with three inputs, two hidden units and two outputs.

hidden layer since it is in some sense hidden between the external inputs (φ_1, φ_2, φ_3) and the output layer. A three-layer network will thus have two hidden layers and so forth. The depicted network is said to be *fully connected* since all inputs/all units in one layer are connected to all units in the following layer.

The mathematical formula expressing what is going on in the MLP-network takes the form

$$\hat{y}_i(t) = g_i[\varphi, \theta] = F_i \left[\sum_{j=1}^{n_h} W_{i,j} f_j \left(\sum_{l=1}^{n_\varphi} w_{j,l}\varphi_l + w_{j,0} \right) + W_{i,0} \right] . \qquad (1.1)$$

θ specifies the *parameter vector*, which contains all the adjustable parameters of the network; i.e., the weights and biases $\{w_{j,l}, W_{i,j}\}$. Since the bias can be interpreted as a weight acting on an input clamped to 1, the joint description "weight" will most often be applied covering both weights and biases.

To determine the weight values one must have a set of examples of how the outputs, $\hat{y}_i$, should relate to the inputs, φ_l. The task of determining the weights from these examples is called *training* or *learning*, and it is basically a conventional estimation problem. That is, the weights are estimated from the examples in such a way that the network, according to some metric, models the true relationship as accurately as possible.

1.2.3 Choice of Neural Network Architecture

Before the training can be performed, some issues need special attention. Unfortunately, not all questions are easily answered:

- What type of relationships can be learned with a multilayer perceptron?
- How many hidden layers should the network have and how many units should be included in each layer?
- How should the activation functions be chosen?

In Cybenko (1989) it is shown that all continuous functions can be approximated to any desired accuracy, in terms of the uniform norm, with a network of one hidden layer of sigmoidal (or hyperbolic tangent) hidden units and a layer of linear output units. The paper does not explain how many units to include in the hidden layer. This issue is addressed in Barron (1993) and a significant result is derived about the approximation capabilities of two-layer perceptron networks when the function to be approximated exhibits a certain smoothness. Unfortunately, the result is difficult to apply in practice for selecting the number of hidden units. Just knowing that any continuous function can be approximated reasonably well by an MLP-network is, however, reassuring since this covers most functions of practical interest.

Due to the above mentioned results one might think that there is no need for using more than one hidden layer and/or mixing different types of activation functions. This is not quite true as it may occur that accuracy can be improved using a more sophisticated network architecture. In particular when the complexity of the mapping to be learned is high, it is likely that the performance can be improved. However, since implementation, training, and statistical analysis of the network become more complicated, it is customary to apply only a single hidden layer of similar activation functions and an output layer of linear units.

1.2.4 Models of Dynamic Systems

The MLP-network is straightforward to employ for discrete-time modelling of dynamic systems for which there is a nonlinear relationship between the system's input and output:

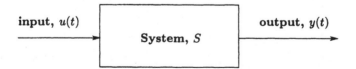

Figure 1.6. A dynamic system with one input (u) and one output (y).

Let t count the multiple of sampling periods so that $y(t)$ specifies the present output while $y(t-1)$ signifies the output observed at the previous sampling instant, etc. If it is assumed that the output of the dynamic system at discrete time instances can be described as a function of a number of past inputs and outputs

$$y(t) = S\left[y(t-1), \dots, y(t-n), u(t-1), \dots, u(t-m)\right] \qquad (1.2)$$

an MLP-network can be used for approximating S if the inputs to the network $(\varphi_1, \varphi_2, \dots)$ are chosen as the n past outputs and the m past inputs:

$$\hat{y}(t|\theta) = g[\theta, \varphi(t)] = \sum_{j=1}^{n_h} W_j f_j \left[\sum_{l=1}^{n+m} w_{j,l}\varphi_l(t) + w_{j,0}\right] + W_{i,0}. \qquad (1.3)$$

This is just a very simple example of the principle for modelling nonlinear dynamic systems with multilayer perceptrons. Models of much greater sophistication can be introduced if, for example, the system is affected by noise or if state-space models are applied. One of the attractive features of using MLP-networks to model unknown nonlinear systems is that a separate discretization process is rendered unnecessary since a nonlinear discrete model is available immediately.

1.2.5 Recurrent Networks

The perceptron network need not have the exact feedforward structure shown in Figure 1.5. In fact, in the context of models for dynamic systems it is often seen that the MLP-network architecture is augmented with feedback loops as illustrated in Figure 1.7. In this case the network is referred to as a *recurrent network*. Unlike the feedforward network, where there is an algebraic relationship between input and output, the recurrent architecture contains memory, i.e., it is a dynamic system. The recurrent network contains the feedforward network as a special case and obviously it therefore represents a more general class of architectures. The mathematical expression governing the network in Figure 1.7 is given by

$$\hat{y}(t|\theta) = g_i[\theta, \varphi(t), t]$$

$$= F_i \left[\sum_{j=1}^{n_h} W_{i,j} f_i(\bullet, t) + W_{i,0}\right] \qquad (1.4)$$

$$= F_i \left[\sum_{j=1}^{n_h} W_{i,j} f_i \left(\sum_{l=1}^{n+m} w_{j,l}\varphi_l(t) + \sum_{l=1}^{n_h} \omega_{j,l} f_l(\bullet, t-1) + w_{j,0}\right) + W_{i,0}\right]$$

The recurrence can be implemented in many different ways. The example shown in Figure 1.7 is just one example. If also the output of the output units are fed back, the network is often said to be *fully recurrent*. The recurrent networks considered later in this book do not have the internal feedback but only a feedback from output to input.

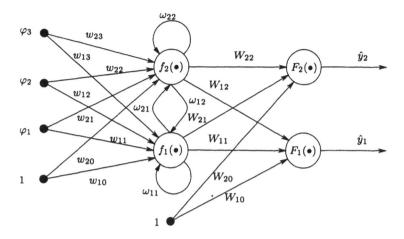

Figure 1.7. A simple example of a recurrent network. The output of the hidden units are fed back as inputs to the network. The loops contain a time delay of the signals they feed back.

1.2.6 Other Neural Network Architectures

The multilayer perceptron is just one out of many neural networks, but it is the only one covered here. Several textbooks are available that offer introductions to the most common network types. Most of these books also provide the historical background for neural networks and how they relate to biological neural networks. See for example Hertz et al. (1991), Haykin (1998), and Zurada (1992). Not all networks are equally suitable for modelling and control of dynamic systems. For these applications the most common alternative to the multilayer perceptron network is probably the Radial Basis Function (RBF) networks (Sanner and Slotine, 1992; Tzirkel-Hancock and Fallside, 1992). Studying and comparing the performance of different network types is however beyond the scope of this book. Nevertheless, much of the theory discussed in the book applies more or less directly to other types of neural networks as well.

2. System Identification with Neural Networks

System identification is the task of inferring a mathematical description, *a model*, of a dynamic system from a series of measurements on the system. There can be several motives for establishing mathematical descriptions of dynamic systems. Typical applications encompass simulation, prediction, fault detection, and control system design. If the burden associated with building a model using laws of physics, chemistry, economics, etc., is considered overwhelming, system identification techniques are naturally of particular interest.

In this chapter the attention is drawn to identification of neural network models for nonlinear dynamic systems. Apart from an increase in complexity compared to identification of linear systems, many of the results known from conventional system identification apply to neural-network-based identification as well. In this chapter, a generic working procedure will be developed to guide the user all the way from an initial experiment to a good model of the system under consideration. This working procedure includes a number of steps that are each implemented to achieve a high degree of automation, numerical reliability, and computational efficiency.

2.1 Introduction to System Identification

Depending on the level of *a priori* insight about the system, the identification problem can be approached in different ways. If the identification is based exclusively on measured data, assuming no or only diminutive knowledge about the physics of the system, the identification process is called *black-box modelling*. In contrast to this, the phrase *white-box modelling*! is used for a pure physical modelling of the system. When a certain level of insight about the system exists and is utilized to improve the empirical modelling, the phrase *gray-box modelling* is used.

This chapter deals primarily with black-box modelling. Mainly because it is difficult to give a generic treatment of gray-box modelling for general nonlinear systems. It is simply impossible to embrace all types of *a priori* knowledge that might exist. Another problem is that even though knowledge about the system is available, it is difficult to utilize it as it relates to a continuous-time description of the system (i.e., in terms of differential equations). Transferring the knowledge to a discrete-time description is often hard and thus it is frequently lost in the discretization process. Confining oneself to black-box modelling is therefore not uncommon regardless that a certain level of insight is available. However, a fundamental understanding of the system's behavior is always useful and facilitates the identification. Such insights can include: the order of the system, whether the dynamics are slow or fast, adequate sampling frequency, stability properties, operating range, time delay, degree of nonlinearity (is it almost linear?), and basic characteristics of the nonlinearities (hard/smooth).

Regardless of the fact that all systems in principle are nonlinear, the major part of the literature on system identification deals with identification of linear systems. There are many reasons for this. Some of the most important are:

- Many systems can be described well by a linear model and when it is not reasonable to use a single linear model for the entire range of operation, it is sometimes possible to identify different linear models for different regimes of the operating range.

- From a computational perspective it is less complicated to perform.

- The analysis is less complicated from a statistical perspective.

- It is *much* simpler to design controllers for a system described by a linear model.

Naturally, the last item on the list is of particular interest in this context. However, the nonlinearities are frequently of such a severe character that employing a nonlinear model in the control design can enhance the performance of the control system greatly. Obviously, nonlinear black-box modelling is therefore important in relation to control of unknown nonlinear systems. This chapter is devoted to explore ways of using neural networks in black-box identification while the subsequent chapter, Chapter 3, will deal with the design of control systems. Since the interest in system identification goes far beyond the application to control, the subject will be treated in a fairly general fashion. Thus, the reader need not be particularly interested in the control engineering perspectives to benefit from studying this chapter.

This chapter attempts to provide a continuous presentation of techniques that together constitute a practical procedure for system identification with neural networks. This includes a selection of material from conventional system identification, optimization theory, nonlinear regression, and the theory of neural networks. Neural-network-based system identification can be implemented as a natural extension to conventional identification and the approach taken in the chapter is thus very much along the lines of the authoritative treatments of linear system identification found in for example Ljung (1999) and Söderström and Stoica (1989). An attempt has been made to describe the implementation aspects with reasonable detail in order to enable the reader to implement all the described methods.

2.1.1 The Procedure

When attempting to identify a model of a dynamic system it is common practice to follow the procedure depicted in Figure 2.1.

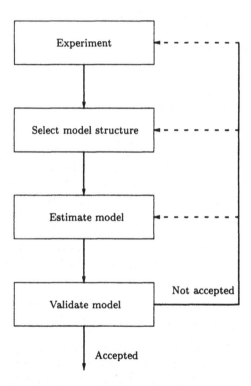

Figure 2.1. The basic system identification procedure.

Naturally, issues like physical insight and intended use of the model will influence all stages in the procedure. A preliminary discussion of the implementation of each stage as well as the route through the diagram is given in the following.

Experiment. The purpose of the experiment is to collect a set of data that describes how the system behaves over its entire range of operation. The idea is too vary the input(s), u, and observe the impact on the output(s), y (see Figure 1.6).

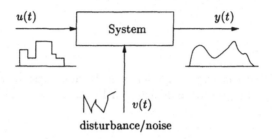

Figure 2.2. An input is applied to the system and it is observed how the output is affected.

The data set of corresponding inputs and outputs

$$Z^N = \{ \, [u(t), y(t)], \; T = 1, \; ... \, , N \}$$

is later used for inferring a model of the system. If the system to be identified is unstable or contains lightly damped dynamics it may be necessary to conduct the experiment in closed-loop. Either by introduction of a manually tuned, stabilizing feedback controller or by using a human operator for controlling the system. Some of the main issues in the experiment stage are: choice of sampling frequency, design of a suitable input signal, and preprocessing of data. Data preprocessing includes, e.g., nonlinearity tests and removal of disturbances, noise, and other undesired effects from the data.

Model structure selection. A *model structure* is a set of candidate models. That is, a set inside which one should search for a model. On a general level the problem of selecting a model structure is twofold:

1. Select a "family" of model structures considered appropriate for describing the system, e.g., linear model structures, multilayer perceptron networks, radial basis function networks, wavelets, or Hammerstein models.

2. Select a subset of the chosen family of model structures. In the family of linear structures this can for instance be an ARX(2,3,1) model structure where (2,3,1) signifies a time delay of one sampling period and that the present output depends on two past outputs and three past inputs.

Both input-output models and state space models will be considered, although most attention will be given to the former. The chapter will primarily deal with systems that have only one output, but some guidelines for how to treat multi-output systems will be given as well.

Estimate model. Once a set of candidate models has been chosen, the next step is to pick one particular model from this set. One will typically pick the model that performs best according to some type of criterion. This criterion can be formulated in many different ways but should ideally relate to the intended use of the model. The most common strategy is to pick the model that provides the best one-step ahead predictions in terms of the smallest expected squared error between observed outputs and predictions. The process of picking a model from the model structure is in the statistical literature known as *estimation*. However, for historical reasons the same process in the neural network community is usually called *training* or *learning*.

Validation. When a model has been estimated/trained it must be evaluated to investigate whether or not it meets the necessary requirements. The validation is closely connected to the intended use of the model. It is often the most hand-waved stage in the identification procedure because the requirements for acceptance often are somewhat fuzzy.

Going backwards in the procedure. The paths going from the validation block and back to the previous stages indicate that the procedure is executed in an iterative manner. It is necessary to go back in the procedure to determine a number of different models, to try out various model structures, and in the worst case even redo the experiment:

- *Path leading back to model estimation:* Although one has a simple criterion characterizing the best model in the model structure it is often difficult to guarantee that the training algorithm will converge to this model. The problem is that a criterion will have several local minima and finding the global minimum is not easy. The feedback path also covers up an augmentation of the criterion called *regularization* or *weight decay*. This extension is very important.

- *Path leading back to model structure selection:* Some strategies are available that to some extent can automate the model structure selection. Perhaps the most popular strategy is called pruning: an initial model structure that is large enough to describe the system is determined and it is then reduced gradually until the optimal structure is achieved.

- *Path leading back to experiment:* If it seems impossible to determine a decent model regardless of how the model structure is selected it may indicate that the data material is insufficient. Either because additional processing, such as a filtering, is needed or because there is simply a lack of information in the data set. The latter is very common and typically implies that certain regimes of the operating range are not reflected in the data set. In this case it is necessary to make an additional experiment to acquire more information about the missing regimes.

First things first - but not necessarily in that order. The chapter is organized as follows: Section 2.2 to Section 2.5 will detail each of the four fundamental steps but for pedagogical reasons in a slightly reversed order. Section 2.2 will discuss model structure selection. On the basis of well-known linear model structures, extensions based on neural networks are proposed that are suitable for nonlinear systems. In Section 2.3 different aspects of the experiment will be treated. Section 2.4 presents training algorithms appropriate in relation to neural-network-based model structures. Particular emphasis will be placed on how to obtain a rapid and robust convergence. The concept of generalization will be introduced and it will be discussed how regularization can be used for tuning a model's ability to generalize to unseen inputs. Section 2.5 will consider different techniques for validating neural network models. Some are based on a statistical foundation while others have a somewhat hand-waved character. The different paths going back in the procedure will be discussed in Section 2.6. Most attention will be paid to pruning algorithms. Finally, Section 2.7 will recapitulate on the overall issues of the chapter and a generic working procedure for system identification with neural networks will be outlined.

2.2 Model Structure Selection

The purpose of this section is to introduce model structures suitable for identification of nonlinear dynamic systems in a stochastic environment. To provide the necessary capability to describe nonlinear mappings, multilayer perceptron neural networks will be used. A number of neural-network-based model structures representing a further development of well-known linear model structures will be proposed. Using linear model structures as the starting point has several advantages. In addition to being a natural approach for those already familiar with conventional identification techniques, it is also convenient in relation to the control system designs discussed in Chapter 3. By way of introduction, this section gives a review of different basic linear model structures. Neural-network-based model structures suitable for identification of nonlinear systems are subsequently introduced as generalizations of the linear model structures.

2.2.1 Some Linear Model Structures

According to Ljung (1999) a system is called linear if it is possible to describe it by a model that takes the form:

$$y(t) = G(q^{-1})u(t) + H(q^{-1})e(t) , \tag{2.1}$$

where G and H are transfer functions in the time delay operator, q^{-1}. The delay operator works on a signal in the following way:

$$q^{-d}x(t) = x(t - d) \tag{2.2}$$

where d is some multiple of the sampling period. $e(t)$ is a white noise signal that is independent of past inputs and that can be characterized by some probability density function. In the multivariable (or MIMO = multi-input, multi-output) case, $u(t)$, $y(t)$, and $e(t)$ are vectors and G and H matrix polynomials in q^{-1}.

If the system is linear the objective of the identification procedure is to determine good estimates of the two transfer functions G and H. The criterion defining the meaning of "good" will in this chapter primarily relate to model's ability to produce one-step ahead predictions with errors of low variance. For the general linear system described by (2.1), it is easily verified that the minimum variance (one-step ahead) prediction is given by:

$$\hat{y}(t|t - 1) = H^{-1}(q^{-1})G(q^{-1})u(t) + \left[1 - H^{-1}(q^{-1})\right] y(t) . \tag{2.3}$$

This alternative representation of the model is occasionally denoted the predictor form of the model.

To clarify the meaning of the phrases *system*, *model structure* and *model* some terminology will be defined in the following (see also Ljung (1999)). Not only will this be of importance in much of the theory derived later in this chapter, it will also help point out the differences in terminology between the system identification and neural network communities.

- The **true system** is assumed to be described by

$$y(t) = G_0(q^{-1})u(t) + H_0(q^{-1})e_0(t) , \tag{2.4}$$

with $e_0(t)$ being a white noise signal independent of the input signal $u(t)$.

- The **model structure**, $\mathcal{M}$, is here a parametrized set of candidate models

$$\mathcal{M} : \left\{G(q^{-1},\theta), \ H(q^{-1},\theta) \mid \theta \in \mathcal{D}_m\right\}$$
$$y(t) = G(q^{-1},\theta)u(t) + H(q^{-1},\theta)e(t) \tag{2.5}$$

where θ denotes the p adjustable parameters, and $\mathcal{D}_m$ is some subset of $\mathbb{R}^p$ inside which the search for a model should be carried out. The model structure in predictor form is obviously given by

$$\hat{y}(t|t-1,\theta) = H^{-1}(q^{-1},\theta)G(q^{-1},\theta)u(t) + \left[1 - H^{-1}(q^{-1},\theta)\right] y(t) . \tag{2.6}$$

Observe the inclusion of θ as an argument implying that the model structure represents *a set* of models. Unless otherwise specified, only one-step ahead predictions are considered. Consequently, the conditioning on $t - 1$ is omitted from this point on for notational convenience.

The model structure will often be written in the alternative form

$$\hat{y}(t|\theta) = \varphi^T(t)\theta \, , \tag{2.7}$$

where θ is the *parameter vector* and φ is the *regression vector*, which contains past inputs, past outputs, or signals derived from the inputs and outputs.

The basic requirement on the set $\mathcal{D}_m$ is (Söderström and Stoica, 1989)

$$\mathcal{D}_m = \left\{ \theta \; \left| \; \begin{array}{l} H^{-1}(q^{-1},\theta)G(q^{-1},\theta) \text{ asymptotically stable} \\ H^{-1}(q^{-1},\theta) \text{ asymptotically stable} \\ G(0,\theta) = 0, \quad H(0,\theta) = 1 \end{array} \right. \right\} , \tag{2.8}$$

but physical insights may of course impose additional constraints on the set. The stability condition obviously guarantees that the predictor remains stable. The assumption $G = 0$ is included to ensure that the predictions depend on past inputs only (i.e., there is no direct link) while $H = 1$ ensures that the predictions only depend on past output measurements.

For much of the theory to come, a basic requirement will be that the model structure is "large enough" to describe the true system. That is,

$$\mathcal{S} \in \mathcal{M} \, . \tag{2.9}$$

- A **model** is simply a particular choice of parameter vector; say, $\theta = \hat{\theta}$.

In most cases, a simpler model structure than the general form in (2.5) is considered. Some important simplifications will be discussed below. The simplifications typically differ from one another by different assumptions about the spectral density of the noise and how the noise is assumed to enter the system. It is natural to rewrite the general model structure as

$$A(q^{-1})y(t) = q^{-d}\frac{B(q^{-1})}{F(q^{-1})}u(t) + \frac{C(q^{-1})}{D(q^{-1})}e(t) \tag{2.10}$$

where

$$\begin{aligned} A(q^{-1}) &= 1 + a_1 q^{-1} + \; \dots \; a_n q^{-n} \\ B(q^{-1}) &= b_0 + b_1 q^{-1} + \; \dots \; b_m q^{-m} \\ C(q^{-1}) &= 1 + c_1 q^{-1} + \; \dots \; c_k q^{-k} \\ D(q^{-1}) &= 1 + d_1 q^{-1} + \; \dots \; d_l q^{-l} \\ F(q^{-1}) &= 1 + f_1 q^{-1} + \; \dots \; f_r q^{-r} \, . \end{aligned} \tag{2.11}$$

The polynomials $A, C, D, E,$ and F are said to be *monic* as the first coefficient is 1.

The Finite Impulse Response model structure (FIR). The simplest type of model structure corresponds to the choice

$$G(q^{-1}, \theta) = q^{-d} B(q^{-1}) \qquad H(q^{-1}, \theta) = 1 \qquad (2.12)$$

in which case the predictor is given by

$$\hat{y}(t|\theta) = q^{-d} B(q^{-1}) u(t) . \qquad (2.13)$$

This can equivalently be expressed in regression form

$$\hat{y}(t|\theta) = \varphi^T(t) \theta , \qquad (2.14)$$

where $\varphi(t)$ is the regression vector defined by

$$\varphi(t) = [u(t-d) \ ... \ u(t-d-m)]^T . \qquad (2.15)$$

Correspondingly, the parameter vector, θ, is composed as follows

$$\theta = [b_0 \ ... \ b_m]^T \qquad (2.16)$$

A system with poles cannot be described exactly by an FIR-model of finite order. However, if the system is stable, and the impulse response decays reasonably fast, the system can often be approximated well by an FIR-model if $B(q^{-1})$ is selected as the first m coefficients of the impulse response.

ARX model structure (AutoRegressive, eXternal input). The ARX model structure corresponds to the choice

$$G(q^{-1}, \theta) = q^{-d} \frac{B(q^{-1})}{A(q^{-1})} \qquad H(q^{-1}, \theta) = \frac{1}{A(q^{-1})} . \qquad (2.17)$$

The predictor thus takes the form

$$\begin{aligned} \hat{y}(t|\theta) &= q^{-d} B(q^{-1}) u(t) + \left[1 - A(q^{-1}) \right] y(t) \\ &= \varphi^T(t) \theta \end{aligned} \qquad (2.18)$$

with

$$\begin{aligned} \varphi(t) &= [y(t-1) \ ... \ y(t-n), \ u(t-d), \ ... \ , \ u(t-d-m)]^T \\ \theta &= [-a_1, \ ... \ -a_n, \ b_0 \ ... \ b_m]^T . \end{aligned} \qquad (2.19)$$

Although G now has poles, there is still only an algebraic relationship between the prediction and the past inputs and measured outputs. Consequently, the predictor will always be stable; even if the system is unstable. This is a very important feature of the ARX model structure.

It should be mentioned that the names *Controlled AutoRegressive model* (CAR), *equation error model*, and *series-parallel model* frequently are used instead of the name ARX.

ARMAX model structure (AutoRegressive, Moving Average, eXternal input). This model structure is even more general than the ARX structure:

$$G(q^{-1}, \theta) = q^{-d} \frac{B(q^{-1})}{A(q^{-1})} \qquad H(q^{-1}, \theta) = \frac{C(q^{-1})}{A(q^{-1})} . \qquad (2.20)$$

The optimal predictor is

$$\hat{y}(t|\theta) = q^{-d} \frac{B(q^{-1})}{C(q^{-1})} u(t) + \left(1 - \frac{A(q^{-1})}{C(q^{-1})} \right) y(t)$$

$$= q^{-d} B(q^{-1}) u(t) + \left[1 - A(q^{-1}) \right] y(t) + \left[C(q^{-1}) - 1 \right] \varepsilon(t, \theta)$$

$$= \varphi^T(t, \theta) \theta . \qquad (2.21)$$

$\varepsilon(t, \theta) = y - \hat{y}(t|\theta)$ represents the *prediction error* or *residual*. The regression and parameter vectors are defined by

$$\varphi(t, \theta) = [y(t-1) \ ... \ y(t-n), \ u(t-d) \ ... \ u(t-d-m),$$

$$\varepsilon(t, \theta), \ ... \ , \varepsilon(t-k, \theta)]^T$$

$$\theta = [-a_1, \ ... \ - a_n, \ b_0 \ ... \ b_m, \ c_1, \ ... \ c_k]^T . \qquad (2.22)$$

Due to the presence of the C-polynomial the predictor now has poles. Thus, C must have its roots inside the unit circle for the predictor to be stable. Also, the poles imply that the regression vector depends on the model parameters, which, as seen later, makes the estimation of model parameters more complicated. Notice that the model dependency was indicated by including θ as an argument to φ in (2.22).

Output error model structure (OE). The output error (or *parallel*) model structure is used if the only noise affecting the system is white measurement noise

$$y(t) = q^{-d} \frac{B(q^{-1})}{F(q^{-1})} u(t) + e(t) , \qquad (2.23)$$

corresponding to the following choice of G and H

$$G(q^{-1}, \theta) = q^{-d} \frac{B(q^{-1})}{F(q^{-1})} \qquad H(q^{-1}, \theta) = 1 . \qquad (2.24)$$

The predictor for this system is simply

$$\hat{y}(t|\theta) = q^{-d} \frac{B(q^{-1})}{F(q^{-1})} u(t)$$

$$= q^{-d} B(q^{-1}) u(t) + \left[1 - F(q^{-1}) \right] \hat{y}(t|\theta) \qquad (2.25)$$

$$= \varphi^T(t, \theta) \theta ,$$

where

$$\varphi(t,\theta) = [\hat{y}(t-1|\theta), \; ... \; , \hat{y}(t-r|\theta), \; u(t-d), \; ... \; u(t-d-m)]$$
$$\theta = [-f_1, \; ... \; , -f_r, \; b_0, \; ... \; , b_m]^T \; . \tag{2.26}$$

For the predictor to be stable, the roots of F must be inside the unit circle.

The State Space Innovations Form (SSIF). The state space description is a widely used alternative to the input-output model structures discussed above. Assume the system can be described by the following set of coupled first-order difference equations

$$x(t+1) = A(\theta)x(t) + B(\theta)u(t) + w(t) \tag{2.27}$$
$$y(t) = C(\theta)x(t) + v(t) \; , \tag{2.28}$$

where $w(t)$ and $v(t)$ are white noise signals independent of the control signal, $u(t)$, and

$$\mathbf{E}\left\{ \begin{bmatrix} w(t) \\ v(t) \end{bmatrix} \begin{bmatrix} w^T(t) \; v^T(t) \end{bmatrix} \right\} = \begin{bmatrix} R_w(\theta) & R_{wv}(\theta) \\ R_{wv}^T(\theta) & R_v(\theta) \end{bmatrix} \; . \tag{2.29}$$

It can be shown (Söderström and Stoica, 1989) that the optimal one-step ahead predictor for this system takes the form

$$\hat{x}(t+1,\theta) = A(\theta)\hat{x}(t,\theta) + B(\theta)u(t) + K(\theta)\varepsilon(t,\theta) \tag{2.30}$$
$$\hat{y}(t|\theta) = C(\theta)\hat{x}(t,\theta) \; . \tag{2.31}$$

$K(\theta)$ is found from (θ is omitted for notational convenience)

$$K = \left[APC^T + R_{wv}\right]\left[CPC + R_v\right]^{-1} \; , \tag{2.32}$$

where $P(\theta)$ represents the positive semi-definite solution to the stationary Ricatti equation

$$P = APA^T + R_w$$
$$\quad - \left[APC^T + R_{wv}\right]\left[CPC^T + R_v\right]^{-1}\left[APC^T + R_{wv}\right]^T \tag{2.33}$$

The optimal predictor is also known as the *Kalman filter* and the matrix $K(\theta)$ is referred to as the Kalman gain. The form of (2.30) is denoted the *state space innovations form*.

A simple relationship between the state space innovations form and the general input-output form exists (Ljung, 1999)

$$G(q^{-1},\theta) = C(\theta)\left[qI - A(\theta)\right]^{-1}B(\theta) \tag{2.34}$$
$$H(q^{-1},\theta) = C(\theta)\left[qI - A(\theta)\right]^{-1}K(\theta) + I \; . \tag{2.35}$$

By some matrix manipulations it can be verified that the poles of the predictor are the eigenvalues of the matrix $A - KC$ (Söderström and Stoica, 1989).
The set $\mathcal{D}_m$ is thus given by

$$\mathcal{D}_m = \{\ \theta\ |\ \text{eig}\,[A(\theta) - K(\theta)C(\theta)]\ \text{inside the unit circle}\ \}\ . \tag{2.36}$$

When estimating state space models, the elements in $K(\theta)$ are typically estimated directly rather than using the detour of estimating the covariance
matrices and solving the Ricatti equation before computing $K(\theta)$.

The selection of a proper parametrization, i.e., the structure of A, B, C,
and K, is a problem that disfavors the innovations form. It is far more involved than the input-output model structures. The problem is that a fully
parametrized model structure, meaning all elements in A, B, C, and K, must
be estimated, is generally not identifiable from a set of input-output data.
This is because such a structure contains more adjustable parameters than
necessary: the same input-output relationship can be described by different
choices of A, B, C, and K. Sometimes the parametrization can be based
on physical insight, which may remedy the problem. However, when taking
a black-box approach some kind of "generic" identifiable parametrization is
required. In the SISO case a number of so-called *canonical forms* exist and
are frequently used for transforming a transfer function description to a state
space description (Kwakernaak and Sivan, 1972). In the MIMO case it is
more complicated. Ljung (1999) proposes MIMO extensions called overlap-
ping forms that are identifiable and thus suitable in a system identification
context. Ljung's guidelines for selecting a parametrization is recapitulated
below (n specifies the model order and n_y the number of outputs):

*"Let $A(\theta)$ initially be a matrix filled with zeros and with ones along the super-
diagonal. Let then row numbers r_1, r_2, ... r_{ny}, where $r_{ny} = n$, be filled with
parameters. Take $r_0 = 0$ and let $C(\theta)$ be filled with zeros, and then let row i
have a one in column $r_{i-1} + 1$. Let $B(\theta)$ and $K(\theta)$ be filled with parameters."*

(Ljung, 1999, Appendix 4A, pp. 128)

The principle is illustrated below for a fifth-order model structure ($n = 5$)
with two outputs ($n_y = 2$) and one input. $\times$ denotes a parameter to be
estimated

$$A(\theta) = \begin{bmatrix} 0 & 1 & 0 & 0 & 0 \\ \times & \times & \times & \times & \times \\ 0 & 0 & 0 & 1 & 0 \\ 0 & 0 & 0 & 0 & 1 \\ \times & \times & \times & \times & \times \end{bmatrix} \qquad B(\theta) = \begin{bmatrix} \times \\ \times \\ \times \\ \times \\ \times \end{bmatrix} \tag{2.37}$$

$$C(\theta) = \begin{bmatrix} 1\,0\,0\,0\,0 \\ 0\,0\,1\,0\,0 \end{bmatrix} \qquad K(\theta) = \begin{bmatrix} \times\ \times \\ \times\ \times \\ \times\ \times \\ \times\ \times \\ \times\ \times \end{bmatrix} . \qquad (2.38)$$

The structural decisions to be made are thus the model order, n, and a set of row indices, $\{r_i\}_{i=1}^{n_y-1}$.

2.2.2 Nonlinear Model Structures Based on Neural Networks

When widening the focus to also include black-box identification of *nonlinear* dynamic systems, the problem of selecting model structures becomes increasingly difficult. In Chapter 1 it was discussed that the multilayer perceptron network (MLP) is good at learning nonlinear relationships from a set of data. Thus, in the pursuit for a family of model structures suitable for identification of nonlinear dynamic systems, it is natural to bring up MLP networks. By making this choice, the model structure selection is basically reduced to dealing with the following two issues:

- Selecting the inputs to the network.

- Selecting an internal network architecture.

An often-used approach is to reuse the input structures from the linear models while letting the internal architecture be feedforward MLP network. This approach has several attractive advantages, e.g.,

- It is a natural extension of the well-known linear model structures

- The internal architecture can be expanded gradually as a higher flexibility is needed to model more complex nonlinear relationships.

- The structural decisions required by the user is reduced to a level that is reasonable to handle.

- Suitable for design of control systems.

Nonlinear counterparts to the linear model structures presented in the previous section are thus obtained by

$$y(t) = g\left[\varphi(t,\theta),\theta\right] + e(t) , \qquad (2.39)$$

or on predictor form

$$\hat{y}(t|\theta) = g\left[\varphi(t,\theta),\theta\right] . \qquad (2.40)$$

$\varphi(t,\theta)$ is again the regression vector while θ is the vector containing the adjustable parameters in the neural network known as the *weights*. g is the function realized by the neural network and it is assumed to have a feed-forward structure. Depending on the choice of regression vector, different nonlinear model structures emerge. If the regression vector is selected as for ARX models, the model structure is called NNARX as the acronym for Neural Network ARX. Likewise, NNFIR, NNARMAX, NNOE, and NNSSIF structures are introduced.

NNFIR and NNARX. As for their linear counterparts, the predictors are always stable because there is a pure algebraic relationship between prediction and past measurements and inputs. This is particularly important in the nonlinear case since the stability issue is here much more complex than for linear systems. The model structures are depicted in Figure 2.3. The absence

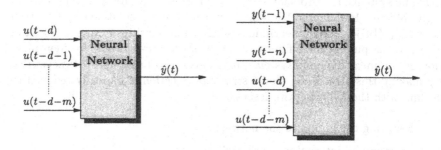

Figure 2.3. The NNFIR (left) and the NNARX (right) model structures.

of stability related problems makes the model structures, in particular the NNARX structure, the preferred choice when the system is deterministic or the noise level is insignificant.

NNARMAX. Although the function g in (2.40) is realized by a feedforward network, the predictor will have a feedback when the regressors are selected as in an ARMAX model

$$\varphi(t, \theta) = [y(t) \ ... \ y(t - n), \ u(t - d) \ ... \ u(t - d - m),$$
$$\varepsilon(t, \theta), \ ... \ , \varepsilon(t - k, \theta)]^T \ . \tag{2.41}$$

The past prediction errors depend on the model output and consequently they establish a feedback. See Figure 2.4. A network model with feedback is usually referred to as a *recurrent network* (Hertz et al., 1991). By investigating the roots of the C-polynomial it is easy to check that the predictor of a linear ARMAX model is stable. For NNARMAX models it is more difficult to analyze the stability properties. Typically it is relevant to somehow consider the stability as a local property. It might be that an NNARMAX model is

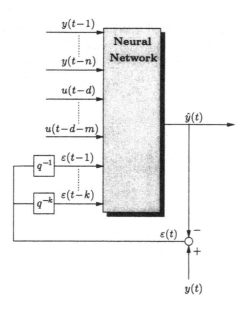

Figure 2.4. The NNARMAX model structure.

stable when operated in some regimes while it is unstable in others. Whether or not this will lead to problems in practice thus depends on how the model is operated.

NNOE. Some of the regressors in the NNOE structure are the predictions of past outputs and it is therefore subject to the same problems as the NN-ARMAX structure. The NNOE model structure is depicted in Figure 2.5.

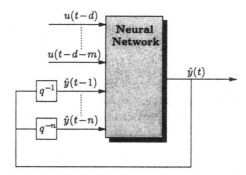

Figure 2.5. The NNOE model structure.

NNSSIF. The difficulty associated with forming a nonlinear extension to the state space innovations form is more pronounced than for the input-output model structures. Here the ideas proposed in Sørensen (1993) and Sørensen (1994) has been adopted. The principle is to let the predictor take the form shown in Figure 2.6.

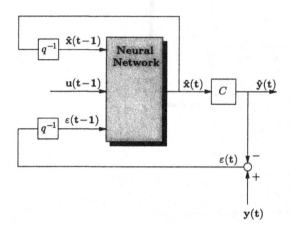

Figure 2.6. The NNSSIF model structure.

$$\hat{x}(t+1|\theta) = g\left[\varphi(t,\theta),\theta\right] \tag{2.42}$$

$$\hat{y}(t|\theta) = C(\theta)\hat{x}(t|\theta), \tag{2.43}$$

where

$$\varphi(t,\theta) = \begin{bmatrix} \hat{x}(t|\theta) \\ u(t) \\ \varepsilon(t,\theta) \end{bmatrix}. \tag{2.44}$$

Like for NNOE and NNARMAX models the predictor contains feedback since the previous state and prediction error vectors both are components of the regression vector.

The issue of identifiability discussed for the linear SSIF model structures is of course carried over to the nonlinear case. One way of dealing with the problem is to introduce two separate networks each responsible for predicting a portion of the states:

Let again $\{r_i\} = \{r_1, r_2, \dots, r_n = n\}$ refer to the states corresponding to the rows in the matrix $A(\theta)$ in (2.30) that are filled with parameters. Moreover, let $\{q_i\} = \{q_1, q_2, \dots, q_{n-n_y}\}$ denote the numbers referring to the remaining states. The two state predictors then take the form

$$\hat{x}_{\{q_j\}}(t+1|\theta) = \hat{x}_{\{q_j+1\}}(t|\theta) + g_q\left[\varphi_q(t,\theta),\theta_q\right] \tag{2.45}$$

$$\hat{x}_{\{r_i\}}(t+1|\theta) = g_r\left[\varphi(t,\theta),\theta_r\right] \tag{2.46}$$

$$\varphi_q^T(t,\theta) = \left[u^T(t)\ \epsilon^T(t,\theta)\right] \ . \tag{2.47}$$

As pointed out in Sørensen (1993), the identified NNSSIF model can be regarded as an *extended Kalman filter* for unknown nonlinear systems. The model structure is thus of potential interest in relation to state space design of controllers. The structural decisions one must make includes: the model order (n), the architecture of the two networks (g_q and g_r), and the indices $\{r_i\}_{i=1}^{n_y-1}$.

Variations and hybrid model structures. It is often proposed to use variations of the above mentioned model structures as well as to create mixtures of linear and nonlinear model structures. Some useful examples are discussed in the following.

- It is sometimes seen that the NNARX model structure is composed of two separate networks or by a network and a linear term (Narendra and Parthasarathy, 1992; Bittani and Piroddi, 1993)

$$\hat{y}(t|\theta) = g_y\left[\varphi_y(t),\theta_y\right] + g_u\left[\varphi_u(t),\theta_u\right] \tag{2.48}$$

$$\hat{y}(t|\theta) = g\left[\varphi_y(t),\theta_y\right] + q^{-d}B(q^{-1})u(t) \tag{2.49}$$

$$\hat{y}(t|\theta) = \left[1 - A(q^{-1})\right]y(t) + g\left[\varphi_u(t),\theta_u\right] \ , \tag{2.50}$$

where

$$\varphi_y(t) = [y(t-1)\ ...\ y(t-n)]^T$$

$$\varphi_u(t) = [u(t-d)\ ...\ u(t-d-m)]^T \ . \tag{2.51}$$

- In a similar manner, various output error model structures can be generated. A modification of (2.50) is particularly interesting

$$\hat{y}(t|\theta) = \left[1 - A(q^{-1})\right]\hat{y}(t,\theta) + g\left[\varphi_u(t),\theta_u\right] \tag{2.52}$$

since this overcomes the problems with unknown stability characteristics of OE-models.

- Following the same logical deduction, a variation of the NNARMAX model structure can be derived as

$$\hat{y}(t|\theta) = g\left[\varphi_{uy}(t),\theta_{uy}\right] + C(q^{-1})\varepsilon(t,\theta) \ , \tag{2.53}$$

where

$$\varphi_{uy}(t) = [y(t-1)\ ...\ y(t-n),\ u(t-d)\ ...\ u(t-d-m)]^T \ . \tag{2.54}$$

- Sometimes model structures that are not motivated by conventional linear model structures can be interesting. As it will appear in Chapter 3 the following model structure is one such example

$$\hat{y}(t|\theta) = g_1\left[\bar{\varphi}(t), \theta_1\right] + g_2\left[\bar{\varphi}(t), \theta_2\right] u(t-1), \qquad (2.55)$$

  where

$$\bar{\varphi}(t) = \left[y(t-1) \ ... \ y(t-n), \ u(t-2) \ ... \ u(t-1-m)\right]^T. \qquad (2.56)$$

- Physical knowledge may obviously influence the choice of regressors. This could imply raising some of the signals to different powers or applying combinations of the different signals.

- It is possible to implement the model structures so that they contain their linear counterpart as a subset. This is obtained by introducing direct connections from inputs to outputs.

- Except for the NNOE structure, it is straightforward to modify the model structures for time series analysis (i.e., no external input). The past inputs are simply omitted in the regression vector. In time series analysis, an experiment is not conducted as it is in system identification since the system cannot be influenced. Apart from the section covering the experiment, all issues discussed in the present chapter are, however, applicable to time series as well.

2.2.3 A Few Remarks on Stability

Stability plays a very important role in control theory. It is a necessary condition for feasibility of the control system that the closed-loop system, consisting of a controller and the system to be controlled, is stable. Additionally, the controller must of course be designed in such a way that the behavior of the closed-loop system satisfies various requirements, e.g., with respect to speed and damping.

Also in system identification one must sometimes deal with the stability issue. It does, however, not play the same vital role as in control theory. In system identification, stability is important in connection to asymptotic analysis of estimation methods. Furthermore, stability is also important in relation to practical implementation of training methods. If, for example, the predictor is unstable for certain choices of model parameters, numerical problems may occur during the training.

In the following a few remarks on stability for discrete nonlinear models will be given. For a more thorough treatment of the topic the reader is referred to the rich literature covering the field; e.g., Salle and Lefschetz (1961), Khalil (1996) or Slotine and Li (1991).

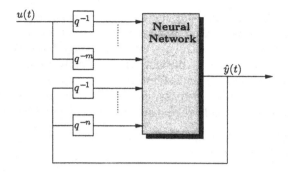

Figure 2.7. Example of a neural network model containing feedback.

The basic block in the model structures described above is in the form

$$y(t) = g\left[\theta, \varphi(t)\right]$$

as depicted in Figure 2.7. This description can quite easily be transformed into the form

$$X(t+1) = F_u\left[X(t), U(t)\right], \qquad X(t_0) = X_0, \qquad (2.57)$$

where $X(t)$ is the part of the regression vector $\varphi(t)$ containing past outputs from the neural network and F_u consist of the neural network function g and the shift function.

Consider in the following a dynamic system described by a discrete-time state space model

$$X(t+1) = F\left[X(t), t\right], \qquad X(t_0) = X_0, \qquad (2.58)$$

where $t \in \mathbb{Z}$. Notice, that (2.57) can be transformed into (2.58) if $U(t)$ is a time function or if $U(t)$ is a state dependent function (or a combination of both). Let $X_o(t)$ be a (nominal) solution to (2.58) (where $X(t_0) = X_0$) and let $x(t) = X(t) - X_o(t)$. Then the deviation from the nominal solution can be described by

$$x(t+1) = f(x(t), t), \qquad (2.59)$$

where the nominal solution corresponds to $x(t_0) = 0$ and where

$$0 = f(0, t) \ \forall t \geq t_0.$$

Stability of the nominal solution can now be defined as

Definition 2.2.1 (Stability) *The solution $x(t) = 0$ is a stable solution to (2.59) if*

$$\forall \epsilon > 0 \ \exists \delta(\epsilon, t_0) > 0 : \qquad \|x(t_0)\| \leq \delta \ \Rightarrow \ \|x(t)\| \leq \epsilon \ \forall \ t \geq t_0. \quad (2.60)$$

$\square$

This is only valid at the sampling instances; it does not say anything about what happens in between. This is also what is most interesting in system identification. However, in control system design the inter-sample behavior is very important as the system to be controlled generally is continuous. In most practical situations a stability in discrete-time will also imply continuous-time stability, but it need not be the case, and thus a more rigorous treatment of the stability issue will usually take place in a continuous-time framework.

If the statement in (2.60) is valid for a δ independent of t_0, then the solution is said to be *uniformly stable*.

Notice that the stability of a solution is a continuity property of the solution with respect to the initial state. Stability means that it is possible to get a solution arbitrarily close to the nominal solution by starting at an initial state close enough to the nominal one ($x_0 = 0$). Often one is not satisfied by the fact that the two solutions are arbitrarily close. Asymptotically, the two solutions are also required to be identical.

Definition 2.2.2 (Asymptotic stability) *The solution $x(t) = 0$ is an asymptotically stable solution if it is stable and there exists a $\rho(t_0) > 0$ such that*

$$\|x(t)\| \to 0 \ \text{for} \ t \to \infty \tag{2.61}$$

for all $\|x(t_0)\| \le \rho$. □

Definition 2.2.3 (Exponential stability) *The solution $x(t) = 0$ is exponentially stable if there exists a $c > 0$ and a $\lambda > 0$ such that*

$$\|x(t)\| \le c\|x(t_0)\| \exp(-\lambda(t - t_0))$$

for all $t > t_0$. □

Sometimes stability of the solution is not an issue, what is important to get a bounded output if the input is bounded. Examples of norms used in these cases are

$$\|x(t)\|_{\mathcal{L}_\infty} = \sup_{t \ge 0} \|x(t)\| , \tag{2.62}$$

$$\|x(t)\|_{\mathcal{L}_2} = \left(\sum_{t=0}^{\infty} x^T(t)x(t) \right)^{\frac{1}{2}} , \tag{2.63}$$

$$\|x(t)\|_{\mathcal{L}_p} = \left(\sum_{t=0}^{\infty} \|x(t)\|^p \right)^{\frac{1}{p}} . \tag{2.64}$$

Definition 2.2.4 (BIBO stability) *The solution $x(t) = 0$ is BIBO stable if there exists constants, c_u, c_x such that*

$$\|u(t)\|_{\mathcal{L}} < c_u \qquad \Rightarrow \qquad \|x(t)\|_{\mathcal{L}} < c_x .$$

$\square$

In a neural network model with only bounded activation functions, such as hyperbolic tangents, the output will always be bounded. This will also be the case if there is no connection from input space ($u(t)$ and states $x(t)$) to output consisting of only unbounded (e.g., linear) activation functions.

For linear time invariant systems it is well-known that asymptotic, BIBO, and exponential stability is obtained if the eigenvalues of the system matrix (corresponding to the poles of the transfer function) are strictly inside the stability area, i.e., the unit circle. For autonomous nonlinear systems the origin is asymptotically locally stable if the linearized system matrix has eigenvalues strictly less than one. In connection to time varying non-linear systems and models, such as neural network models with time-varying inputs, the analysis is much more complicated. Under the assumption that the time-variation is sufficiently slow the stability of the neural network or the control system can be established by evaluating the eigenvalues of the linearized system.

In the following, stability is to be understood in more heuristic terms. What is usually meant is that the signals should not "explode" and deviate to far from the desired. Typical stability problems will be that the network or system output goes into excessive oscillations instead of being smooth, or the signals will saturate.

2.2.4 Terminology

A different terminology is often applied when neural networks are used for modelling purposes. This must be attributed to the fact that the neural network field originated more or less independently of the field of statistical modelling. It has already been mentioned that *estimation* is called *training* or *learning*. In both Ljung and Sjöberg (1992) and Sarle (1994) small dictionaries have been made containing "translations" of common expressions. In relation to the previous subsection, a difference between the two fields are the expressions *teacher* and *student*: The true system,

$$y(t) = g_0 \left[\varphi(t, \theta_0), \theta_0 \right] + e(t) , \qquad (2.65)$$

is usually called the *teacher* and the model structure (the set of candidate models),

$$\mathcal{M} : \{ \, g \, [\varphi(t,\theta),\theta] \,\, | \, \theta \in \mathcal{D}_m \subset \mathbb{R}^p \, \}$$

$$y(t) = g \, [\varphi(t,\theta),\theta] + e(t) \, , \tag{2.66}$$

is called the *student space*. A particular model is then called a student. The condition $\mathcal{S} \in \mathcal{M}$ is thus explained verbally as "the teacher lies in the student space".

Other expressions common to the neural network community will be explained whenever it fits in. In this book an attempt has been made to use the well-known system identification terminology defined in, e.g., Ljung (1999). Nevertheless, certain expressions (like training) will appear regularly, having obtained such wide acceptance that it would lead to much confusion if they were not used.

2.2.5 Selecting the Lag Space

The model structure selection is here treated at a quite general level. A more detailed discussion will appear later. Often it is of little importance that the network architecture has been selected a few parameters too small or too large. However, a wrong choice of lag space, i.e., the number of delayed signals used as regressors, may have a disastrous impact on some control applications. Too small obviously implies that essential dynamics will not be modelled but too large can also be a problem. From the theory on linear systems it is known that too large a lag space may manifest itself as common factors in the identified transfer function(s). An equivalent behavior must be expected in the nonlinear case. Although it is not always a problem, common factors (corresponding to hidden modes) will lead to difficulties in some of the controller designs discussed in Chapter 3.

The specific choice of network architecture and regressors will be treated later. Some methods, known as pruning algorithms, can take a model structure which is too large and reduce it gradually until an optimal structure is obtained. However, generating this sufficiently large initial network model may be hard to do on a trial-and-error basis. It is necessary to determine both a sufficiently large lag space and an adequate number of hidden units. While it is difficult to apply physical insight towards the determination of the number of hidden units, it can often dictate the proper lag space. If the lag space is properly determined, the model structure selection problem is substantially reduced. If one has no idea regarding the lag space it is sometimes possible to determine it empirically. In He and Asada (1993) one such method is described, which can be used for deterministic systems. It is based on a prior assumption that the system can be represented accurately by a function that is reasonably smooth in the regressors.

Assume that it is possible to describe the system by a noise-free NNARX model

$$y(t) = g_0\left[\varphi(t), \theta\right]$$
$$\varphi^T(t) = [\varphi_1,\ \varphi_2,\ \dots,\varphi_z,] \tag{2.67}$$
$$= [y(t-1)\ \dots\ y(t-n),\ u(t-d)\ \dots\ u(t-d-m)]\ . \tag{2.68}$$

Let a data set consisting of N input-output pairs be provided

$$Z^N = \{\ [\varphi(t), y(t)],\ t = 1,\ \dots\ N\}\ . \tag{2.69}$$

In Section 2.3 it is explained how to conduct an experiment to acquire such a data set.

Assume the magnitude of the derivative of the system with respect to each of the regressors is bounded by some positive value, B,

$$|g_l| = \left|\frac{\partial g_0}{\partial \varphi_l}\right| \le B \quad l = 1, 2,\ \dots,z\ . \tag{2.70}$$

For all combinations of input-output pairs, the Lipschitz quotient is now introduced:

$$q_{ij} = \left|\frac{y(t_i) - y(t_j)}{\varphi(t_i) - \varphi(t_j)}\right|,\quad i \ne j\ , \tag{2.71}$$

where $|\ |$ specifies the Euclidean norm, i.e., the distance. The Lipschitz condition then states that q_{ij} is always bounded, if the function g_0 is continuous, that is, $0 \le_{ij} \le L$.

Consider now the differences: $\delta y = y(t_i) - y(t_j)$, $\delta\varphi_l = \varphi_l(t_i) - \varphi_l(t_j)$. If the differences $\delta\varphi_l$ are small, the following approximation is fair

$$\delta y = \frac{\partial g}{\partial \varphi_1}\delta\varphi_1 + \frac{\partial g}{\partial \varphi_2}\delta\varphi_2 + \cdots + \frac{\partial g}{\partial \varphi_z}\delta\varphi_z$$
$$= g_1\delta\varphi_1 + g_2\delta\varphi_2 + \cdots + g_z\delta\varphi_z\ . \tag{2.72}$$

Consequently, the Lipschitz quotient must obey

$$q_{ij}^{(z)} = \frac{|\delta y|}{\sqrt{(\delta\varphi_1)^2 + \dots + (\delta\varphi_z)^2}}$$
$$= \frac{|g_1\delta\varphi_1 + \dots + g_z\delta\varphi_z|}{\sqrt{(\delta\varphi_1)^2 + \dots + (\delta\varphi_z)^2}} \le \sqrt{z}B\ . \tag{2.73}$$

The superscript $^{(z)}$ refers to the total number of regressors.

It is interesting to apply this inequality on two different cases: the regression vector is insufficient and the regression vector is too large.

Insufficient number of regressors: Assume that the zth regressor is missing

$$q_{ij}^{(z-1)} = \frac{|\delta y|}{\sqrt{(\delta\varphi_1)^2 + \dots + (\delta\varphi_{z-1})^2}}$$
$$= \frac{\sqrt{(\delta\varphi_1)^2 + \dots + (\delta\varphi_z)^2}}{\sqrt{(\delta\varphi_1)^2 + \dots + (\delta\varphi_{z-1})^2}} \times \frac{|g_1\delta\varphi_1 + \dots + g_z\delta\varphi_z|}{\sqrt{(\delta\varphi_1)^2 + \dots + (\delta\varphi_z)^2}} . \tag{2.74}$$

As an extreme example, imagine that the differences $\delta\varphi_l = 0$ for all l except for $l = z$. If the output depends on the zth regressor, there will obviously be points were the difference $\delta y \neq 0$. Disregarding the regressor φ_z will thus lead to an infinite Lipschitz quotient. In general one cannot rely on the possibility that the data material contains such examples of course. However, it must be expected that the lack of a regressor most often will lead to very large quotients. Moreover, as more regressors are missing, the quotient will increase significantly.

More regressors than necessary: If too many past signals are included in the regression vector the vector will contain redundant information. Consider for instance the case where one additional regressor is included

$$
\begin{aligned}
q_{ij}^{(z+1)} &= \frac{|\delta y|}{\sqrt{(\delta\varphi_1)^2 + \ldots + (\delta\varphi_{z+1})^2}} \\
&= \frac{\sqrt{(\delta\varphi_1)^2 + \ldots + (\delta\varphi_z)^2}}{\sqrt{(\delta\varphi_1)^2 + \ldots + (\delta\varphi_{z+1})^2}} \times \frac{|g_1\delta\varphi_1 + \ldots + g_z\delta\varphi_z|}{\sqrt{(\delta\varphi_1)^2 + \ldots + (\delta\varphi_z)^2}} \quad .(2.75)
\end{aligned}
$$

It is quite obvious that the superfluous regressor will have only a minor impact on the Lipschitz quotient in that it typically will lead to an insignificant reduction.

In the paper of He and Asada (1993) these properties have been used for a criterion for determination of the optimal regressor structure. The entire procedure is outlined below:

1. For a given choice of lag space, determine the Lipschitz quotients for all combinations of input-output pairs.

2. Select the p largest quotients, $p = 0.01N \sim 0.02N$. The largest quotients typically occur when the differences $\delta\varphi_l$ are small.

3. Evaluate the criterion

$$
\bar{q}^{(n)} = \left(\prod_{k=1}^{p} \sqrt{n} q^{(n)}(k) \right)^{\frac{1}{p}} . \tag{2.76}
$$

4. Repeat the calculations for a number of different lag structures.

5. Plot the criterion as a function of lag space and select the optimal number of regressors as the "knee-point" of the curve.

It is very time-consuming to compute all the quotients. In particular if N is large, and one wishes to explore a large number of lag structures. Therefore one might consider letting the number of past inputs and outputs be increased simultaneously. As long as they are not both too large, "common factors" are prevented.

Figure 2.8 illustrates the method applied to a data set obtained from a simulated experiment with a nonlinear dynamic system. The number of past inputs and outputs are increased simultaneously from $n = m = 1$ to $n = m = 8$.

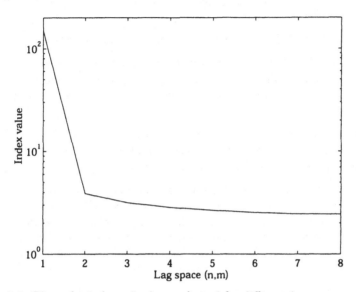

Figure 2.8. The order index criterion evaluated for different lag spaces.

It is seen that the criterion recommends using the lag space $n = m = 2$, which is in fact the correct dimension.

2.2.6 Section Summary

In this section various model structures were presented, including linear and nonlinear neural-network-based model structures.

Model structure. The model structure is a *set* of candidate models. Selecting a model structure implies selecting a set of inputs (regressors) and specifying how to combine the regressors into a one-step ahead prediction.

Linear or nonlinear. If the prediction is obtained by a linear combination of the regressors, the model structure is said to be *linear*. If the regressors are used as input to a neural network, the model structure is *nonlinear*.

Regression vector. The regressors are a number of past signals. Some commonly model structures are obtained by the following choice of regressors:

- FIR/NNFIR: past control inputs.

- ARX/NNARX: past control inputs and observed outputs.

- OE/NNOE: past control inputs and output predictions.

- ARMAX/NNARMAX: past control inputs, output predictions, and residuals.

- SSIF/NNSSIF: past input, state estimate, and residual.

Sometimes hybrids of linear and nonlinear model structures are used. Physical insight might also suggest that regressors are simple functions of signals; i.e., two signals are multiplied or a signal is raised to the power of two.

Network architecture. For nonlinear model structures one has to select an appropriate network architecture in addition to the regressors. This implies specifying a sufficient number of hidden units.

Stability. The predictors must be stable. Instability might occur for model structures that have feedback. NNFIR and NNARX model structures are usually preferred as a first choice as their predictors cannot become unstable. Not even if the underlying system is unstable. Although a predictor in principle should be stable, stability problems might occur during training; or afterwards if the obtained model is inaccurate.

Lag space. In the noise-free case it is sometimes possible to determine the lag space (number of past inputs and outputs) automatically with the so-called Lipschitz coefficients.

Terminology. The same things have different names in different communities. For instance, what is called *training* or *learning* in most neural network literature is called *estimation* in the statistical literature. There are several such differences in terminology.

2.3 Experiment

The primary purpose of an experiment is to produce a set of examples of how the dynamic system to be identified responds to various control inputs. It is the first stage in the identification procedure and thus it constitutes a bottleneck as to how accurate a model can be obtained.

The experiment is particularly important in relation to the *nonlinear* black-box modelling considered in this book. One must therefore be extremely careful to collect a set of data that describes how the system behaves over its entire range of operation.

This section draws attention to various issues in the experiment design, which includes more than just generation of data for neural network training. The following topics will be covered:

- Tests for resolving whether a nonlinear approach is relevant at all.
- Design of input signals leading to an adequately informative data set.
- Techniques for preparing the data for neural network modelling.

Although the choice and location of sensors also could be considered part of the experiment, it is here assumed that the complete system, including sensors, is given. A more or less unlimited control over the input(s) is also assumed. In practice this is not always reasonable of course. For many industrial systems that one may wish to identify, a product is manufactured. Examples are furnaces, reactors, and distillation columns. Varying the input(s) regardless of the production taking place can in these cases be a costly affair. However, strategies for taking such constraints into account are hard to treat generically, and it is considered beyond the scope of this book. In a practical situation, one should thus aim at conducting the experiment in accordance with the guidelines suggested in the following to the extent that no harm is done to the system.

2.3.1 When is a Linear Model Insufficient?

When the objective of the system identification is to produce a model that can be deployed for control system design, one is often willing to accept a quite inaccurate linear model even if the system to be controlled is nonlinear. The reason for this being that it is considerably easier to design and implement controllers based on a linear model. If one has insufficient physical insight to decide whether a linear model is feasible, one can simply estimate a number of different linear models to investigate this. This is the safest but of course also most laborious way of checking it. However, to avoid an unnecessary waste of time it is desirable to have a collection of simple tests at disposal for determining "how linear" the system in question actually is. Below, a few simple *nonlinearity tests* are presented. These and more tests are described in Haber (1985) and Tong (1990).

Superposition check. Two main characteristics of nonlinear systems are that superposition

$$y(t) = g\left[\varphi_1(t) + \varphi_2(t)\right] = g\left[\varphi_1(t)\right] + g\left[\varphi_2(t)\right] \tag{2.77}$$

and homogeneity

$$y(t) = g\left[\alpha\varphi(t)\right] = \alpha g\left[\varphi(t)\right] \tag{2.78}$$

are *not* satisfied. At least not over the entire operating range. If no disturbances affect the system these conditions are often quite easy to check. It is
necessary to assume that the system is at rest for (2.77), (2.78) to hold for
linear systems; otherwise, one must also take into account the initial condition. If the system is stable one can might also wait until the transient has
disappeared before checking the conditions.

As an example, one can try the following procedure: apply a zero input signal
and wait for steady-state to occur to investigate if there is a DC-offset (D).
Do then apply two different input signals, u_1 and u_2, obeying

$$u_2(t) = cu_1(t) \ . \tag{2.79}$$

If the system is linear then the ratio

$$r(t) = \frac{y_2(t) - D}{y_1(t) - D} \tag{2.80}$$

should equal c at all times. As a "non-linearity index" for the system, one
can for example use

$$v = \max_i \left| \frac{r(t) - c}{c} \right| , \tag{2.81}$$

which for linear systems should be zero.

Frequency response check. It is well-known that the frequency response
is unique for linear systems regardless of the amplitude of the input signal.
In order to check for linearity, one can thus apply different sinusoidals to
the system. The frequency as well as the amplitude should be varied. If the
system is linear, the stationary output should be a sinusoidal with the same
frequency, and an amplitude that is proportional to the input. By checking
the presence of subharmonics by a Fourier analysis of the output signal one
can get an idea about the nonlinearity. Naturally, some care should be taken
in case the system is affected by disturbances. Furthermore, if the system
is heavily affected by measurement noise it is recommended that the output
sequence be averaged over a number of trials.

2.3.2 Issues in Experiment Design

If physical knowledge or nonlinearity tests motivates a neural-network-based
identification approach, different issues must be considered in relation to the
acquisition of data appropriate for estimation.

Choice of sampling frequency. If the sampling frequency is selected too high compared to the dynamics of the system under consideration, serious problems with numerical ill-conditioning are likely to occur when attempting to identify a model (Middleton and Goodwin, 1990). If the identification is performed with control system design as the ulterior motive, the selection of sampling frequency thus becomes somewhat involved. In this case, the sampling frequency should also be chosen in accordance with the desired dynamics of the closed-loop system consisting of controller and system. A high sampling frequency permits a rapid reference tracking and a smoother control signal, but the problems with numerical ill-conditioning will become more pronounced. Consequently, the sampling frequency should be selected as a sensible compromise between favoring the identification and favoring the controller design.

The curse of dimensionality. As mentioned above, superposition and homogeneity are not obeyed by nonlinear systems. In turn, this imposes an immense increase in the demand for *excitation* in the input signal. While for identification of linear systems it is sufficient to apply a signal containing a finite number of frequencies, a nonlinear system demands, roughly speaking, that all combinations of frequencies and amplitudes in the system's operating range are represented in the signal. As a consequence, the necessary size of the data set increases dramatically with the number of inputs and outputs. The problem is subject to a so-called *curse of dimensionality* and unfortunately there is no obvious remedy. It is a substantial weakness with nonlinear black-box approaches and essentially it prevents one from deploying neural networks in identification of large scale systems.

Designing the input signal. Before an input signal is selected it is important to identify the operating range of the system. Special care must be taken not to excite dynamics that one does not intend to incorporate in the model (e.g., mechanical resonances). Typically, this implies that the input signal be kept at sufficiently low frequencies. In identification of linear systems it is customary to apply a signal consisting of a number of sinusoids of different amplitude. Also the so-called PRBS signals (Pseudo Random Binary Sequence) are popular. However, when working with non-linear model structures it is important that all amplitudes and frequencies are represented, as mentioned above. Some input signals that attempt to meet this demand are proposed below. For supplementary information see also Söderström and Stoica (1989).

N-samples-constant: Let $e(t)$ be a white noise signal with variance σ_e^2. The signal defined by

$$u(t) = e\left(\text{int}\left[\frac{t-1}{N}\right] + 1\right) \qquad t = 1, 2, \dots \qquad (2.82)$$

will then jump to a new level at each Nth sampling instant (*int* denotes the integer part). Its covariance function is

$$R_u(\tau) = \frac{N - \tau}{N}\sigma_e^2 \, , \tag{2.83}$$

corresponding to the spectral density

$$\phi(\omega) = \frac{\sigma_e^2}{2\pi N} \frac{1 - \cos N\omega}{1 - \cos \omega} \, . \tag{2.84}$$

An example of the signal is depicted in Figure 2.9.

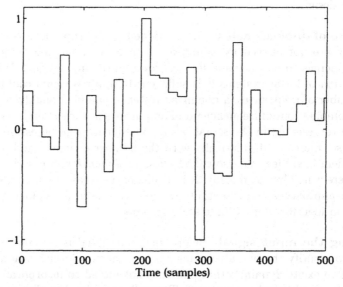

Figure 2.9. The "N-samples-constant" signal for $N = 15$. The signal is scaled to fit the interval $[-1; 1]$.

In most control systems, the controller will produce control inputs that change only a little over two consecutive sampling instants. If $e(t)$ is Gaussian distributed one may therefore consider a "random walk" type of modification, corresponding to a low-pass filtering:

$$u(t) = u(t - N) + e\left(\text{int}\left[\frac{t - 1}{N}\right] + 1 \right) \qquad t = 1, 2, \ \dots \tag{2.85}$$

Level change at random instances: An extension to the N-samples-constant signal is obtained by introducing an additional random variable for deciding when to change the level.

$$u(t) = \begin{cases} u(t-1) & \text{with probability } \alpha \\ e(t) & \text{with probability } 1-\alpha \end{cases} . \qquad (2.86)$$

This signal has the covariance function

$$R_u(\tau) = \alpha^\tau \sigma_e^2 , \qquad (2.87)$$

corresponding to the spectral density

$$\phi(\omega) = \frac{\sigma_e^2}{2\pi} \frac{1-\alpha}{1+\alpha^2 - 2\alpha\cos\omega} . \qquad (2.88)$$

The signal is shown in Figure 2.10.

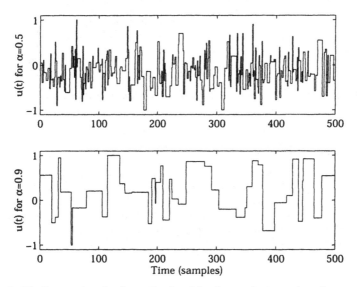

Figure 2.10. Input signal where the level is changed at random instances. The signal is shown for $\alpha = 0.5$ and $\alpha = 0.9$ and is in both cases scaled to stay in the interval $[-1;1]$.

Again, one may consider implementing a random walk modification of the signal or some other type of low-pass filtering.

Chirp signal: A chirp signal is a sinusoid with a gradually increasing frequency. With such a signal one can accurately excite the desired frequency range. A chirp signal can be generated according to the procedure below (Franklin et al., 1998)

$$\omega_t = \omega_{start} + \frac{t}{N}(\omega_{final} - \omega_{start})$$

$$u(t) = u_0 + A\sin(\omega_t t T_s), \qquad t+1,2, \dots \qquad (2.89)$$

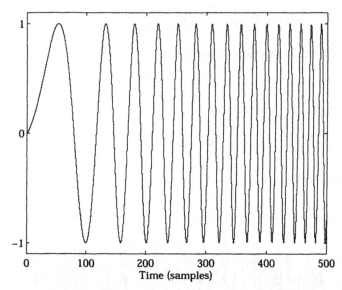

Figure 2.11. A chirp signal. The frequency is increased from $\omega_{start} = 0.01/T_s$ to $\omega_{final} = 0.2/T_s$.

The chirp signal is illustrated in Figure 2.11.

When applying this signal to a nonlinear system it should be repeated for a number of combinations of different DC-values (u_0) and amplitudes (A).

Irrespective of the type of input signal selected, it is important that it conforms to the system in question. There will always be physical limitations that cannot/should not be violated; e.g., constraints are inevitable in D/A-converters and actuators.

Collecting data in closed-loop. When a system is unstable or poorly damped it is recommended (necessary) to use a stabilizing feedback controller to keep the system inside the range in which it is intended to operate. If a stabilizing controller is not available beforehand, one may resort to a manually tuned PID controller or some other simple approach. Sometimes it is also possible to use a human operator for controlling the system. If the model is going to be used for control system design, it is in fact a good idea to collect the data in closed-loop. Preferably with a controller exhibiting a behavior close to what the final controller will show (Gevers, 1993). This is of course a contradiction since if such a controller was available, then it would make no sense to perform the system identification and design a new controller. What one might do in practice is to employ an iterative design.

Identification of systems operating in closed-loop is a subject treated in several text books on identification. In Söderström and Stoica (1989) it is shown that the system's identifiability under certain circumstances can be lost when feedback is introduced. A way to preserve the identifiability is to use either a nonlinear or a time-varying controller. Another way is to add an additional input in the loop (see Figure 2.12)

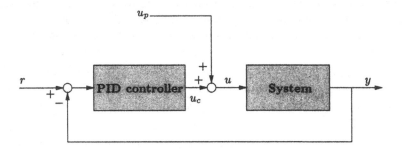

Figure 2.12. A new input signal is added to the output of a manually tuned controller to further excite the system.

Even if the system is in fact identifiable, it is often a good idea to add an additional signal anyhow. A manually tuned controller (or a human operator) will typically perform conservatively in the sense that the input to the system will be a low frequency signal. This may have the effect that the high frequency portion of the operating range is not excited properly. Adding a signal with more energy in higher frequencies may then compensate for this deficiency.

2.3.3 Preparing the Data for Modelling

Intelligent processing of the data is often much more important than trying a large number of different model structures and training schemes. Many different types of processing can be considered for extracting the most valuable information from the measured data and to make it suitable for neural network modelling. Some suggestions are given in the following paragraphs.

Filtering. Filtering is widely used for removing from the measured signal(s) noise, periodic disturbances, off-sets, and the effects of "uninteresting" dynamics. When high frequency noise/disturbances are causing problems, it is recommended to remove it by using an analog pre-sampling filter to avoid aliasing phenomena. Off-set, drift, and low-frequency disturbances can be removed by filtering the data after the sampling.

Remove redundancy and outliers from the data set. Sometimes a large number of input-output pairs from a small regime of the entire operating range dominates the data set. When training on this data set, it is likely that the model obtained will be very accurate in the regime that was over-represented at the expense of a poor performance outside the regime. A little "surgery" on the data set might be necessary here to eliminate redundant information. Apart from obtaining a more equal weighting of the information, a reduction of the data set size also has the benefit that training times will be reduced.

It is also recommended to remove outliers form the data set (or alternatively insert interpolated values of the output signal). With the type of criterion function considered in Section 2.4, outliers will often have a fatal impact on the trained model.

Note that it can be a problem to identify models based on recurrent networks, e.g., NNOE and NNARMAX models, if portions of the data set have been removed. The feedback connections in the model give raise to a transient when an abrupt change in the output signal occurs. These transients will need some time to disappear, and one has to be careful not to let them corrupt the training.

Scaling. It is highly recommended to remove the mean and scale all signals to the same variance. The signals are likely to be measured in different physical units and without scaling there is a tendency that the signal of largest magnitude will be too dominating. Moreover, scaling makes the training algorithm numerically robust and leads to a faster convergence (Le Cun et al., 1991). Finally, experience has shown that it simply tends to give better models.

If the network model is a two-layer network with linear output units, it is straightforward to re-scale the weights after the training session is completed. In this way the final network model can work on unscaled data.

For multi-output systems there might be reasons for scaling the outputs differently if the system is affected by noise. This will be discussed further in Section 2.4.

2.3.4 Section Summary

The primary objective of an experiment is to gather a set of examples of how system responds to various inputs. These examples can later be used for training a neural network to model the system. Another purpose of the experiment is to perform a series of nonlinearity tests to check whether a neural network approach is at all relevant or if a linear approach would be more appropriate. An experiment is usually concluded by cleaning up the collected data to remove undesired properties.

Nonlinearity tests. Prior to collection of data for system identification, it is a good idea to carry out a few nonlinearity tests if the physical insight about the system is limited.

Superposition and homogeneity: Starting from rest, a linear system should be affine in the control input.

Frequency response: If the system stable, application of a sinusoidal on the input should in steady-state result in a sinusoidal on the output with the same frequency.

Experiment design. As it turns out, a surprisingly long time is often spent on the experiment design. Some of the important issues that one has to deal with are:

Sampling frequency: Rarely an easy choice. Not too slow, but not fast either.

Input signal: It is important that the input signal, applied to the system, really "shakes" the system so that all corners of the operating range are covered. Make sure to collect plenty of data, but try to avoid collecting too much redundant information.

Experiment in closed-loop: If the system is poorly damped or unstable, it might be necessary to conduct the experiment while using some type of manually tuned controller.

Preparing data for modelling. The raw data will often contain measurement errors and other information, which one is not interested in incorporating into the model. This should be removed from the data

Filtering: An analog pre-filter should be used prior to the sampling to avoid aliasing. With discrete filtering one can later remove band-limited or low-frequency disturbances.

Outliers: Inspect the data to see if there are obvious outliers. In case there are, they should be removed.

Scaling: Remove the mean from the signal and scale to unit variance. This generally improves the model. Furthermore, the training algorithm will converge more rapidly.

2.4 Determination of the Weights

Assume now that a data set has been acquired and that some model structure has been selected. According to the identification procedure in Figure 2.1, the next step is then to apply the data set to pick "the best" model among the candidates contained in the model structure. This stage is called *training*.

The training can be computationally intensive, but it is generally one of the easiest stages in the identification. It is not very difficult to implement a training algorithm in a computer, but one might as well resort to one of the many available software packages.

The training problem can be rephrased in more formal terms. Given a set of data

$$Z^N = \{ \, [u(t), y(t)], \; T = 1, \, \dots \, , N \}$$

and a set of candidate models

$$y(t) = \hat{y}(t|\theta) + e(t) = g[t, \theta] + e(t)$$

the purpose of the training is to determine a mapping from the data set to the set of candidate models

$$Z^N \rightarrow \hat{\theta} \tag{2.90}$$

so that a model is obtained, which provides predictions that are in some sense close to the true outputs of the system. The most commonly used measure of closeness for this type of problems is specified in terms of a mean square error type of criterion

$$V_N(\theta, Z^N) = \frac{1}{2N} \sum_{t=1}^{N} [y(t) - \hat{y}(t|\theta)]^2 = \frac{1}{2N} \sum_{t=1}^{N} \varepsilon^2(t, \theta) \,. \tag{2.91}$$

This scheme is said to be a *Prediction Error Method (PEM)* (Ljung, 1999) as the objective is to minimize a sum over some norm of the prediction errors. Sometimes other norms than the square are considered to conform to a particular noise distribution or, in the multi-output case, to take into account different noise levels on different outputs. The multi-output case is briefly discussed in Section 2.4.3. With the criterion (2.91), the PEM-method corresponds to maximum likelihood estimation under the assumption that the distribution of the noise signal, $e(t)$, is Gaussian.

The most appealing features of mean square error criteria are the simplicity at which a weight update rule can be derived and that knowledge about the noise distribution generally is of little importance. It may not lead to the perfect model if the noise is not Gaussian, but in many practical situations an accurate model is obtained anyway. For these reasons only mean square error criteria will be considered here. Section 2.4.1 describes different ways to train the network in order to obtain a minimal value of the criterion (2.91). Section 2.4.2 introduces the notion of *average generalization error* and questions the usefulness of the basic mean square error criterion. A small extention to the criterion, known as *regularization*, is suggested to deal with certain common problems. Section 2.4.3 treats different implementation issues relevant to neural network training with PEM-methods.

2.4.1 The Prediction Error Method

In the prediction error method the objective is to determine the weights in the network as the minimizer of the criterion

$$\hat{\theta} = \arg\min_{\theta} V_N(\theta, Z^N) . \qquad (2.92)$$

When the criterion is quadratic in the prediction error, $\varepsilon(t, \theta)$, as in (2.91), training is a so-called *ordinary nonlinear least squares* problem, which is a special case of unconstrained optimization. Nonlinear least squares problems occur within quite diverse fields, and many methods exist for solving them. This section is devoted to presenting some methods considered to be of particular relevance for neural network training.

Searching for a minimum. The second-order Taylor series expansion of the criterion in θ^* is

$$V_N(\theta, Z^N) = V_N(\theta^*, Z^N) + (\theta - \theta^*)^T V_N'(\theta^*, Z^N) \qquad (2.93)$$
$$+ \frac{1}{2}(\theta - \theta^*)^T V_N''(\theta^*, Z^N)(\theta - \theta^*)$$

where the gradient is defined by

$$G(\theta^*) = V_N'(\theta^*, Z^N) = \left.\frac{dV_N(\theta, Z^N)}{d\theta}\right|_{\theta=\theta^*} , \qquad (2.94)$$

and the second-order derivative matrix, the Hessian, is defined by

$$H(\theta^*) = V_N''(\theta^*, Z^N) = \left.\frac{d^2V_N(\theta, Z^N)}{d\theta^2}\right|_{\theta=\theta^*} . \qquad (2.95)$$

Notice that the informal notation $d\theta^2$ has been used for the matrix derivative $d\theta\, d\theta^T$.

A sufficient condition for $\theta = \theta^*$ being a minimum of $V_N(\theta, Z^N)$ is that the gradient equals zero

$$G(\theta^*) = 0 \qquad (2.96)$$

and that the Hessian matrix is positive definite, i.e., that

$$v^T H(\theta^*) v > 0 ., \qquad (2.97)$$

for all non-zero vectors v. The search for a minimum is started out from an initial guess on the parameters, $\theta^{(0)}$. The weights are then adjusted according to some training method. When the criterion, as here, is a complex nonlinear function of the adjustable parameters, the minimum is usually found using an iterative search method. This type of method generally takes the form

$$\theta^{(i+1)} = \theta^{(i)} + \mu^{(i)} f^{(i)} , \tag{2.98}$$

where $\theta^{(i)}$ specifies the current iterate (number i), $f^{(i)}$ the *search direction*, and $\mu^{(i)}$ the *step size*. The iterations are continued until $\theta^{(i)}$ is believed to be sufficiently close to the minimum, $\hat{\theta}$.

The criterion will in general have more than one minimum, but unfortunately search methods of the above kind do not guarantee convergence to the *global* minimum. Which minimum is actually reached depends upon the choice of initial weights $\theta^{(0)}$. See also Section 2.6.1.

In the following, some examples of iterative search methods will be presented:

The gradient method. The principle of the gradient (descent) method, or *steepest descent* as it is sometimes called, is at each iteration to modify the weights along the opposite direction of the gradient. That is, the search direction is selected as $f^{(i)} = -G(\theta^{(i)})$

$$\theta^{(i+1)} = \theta^{(i)} - \mu^{(i)} G(\theta^{(i)}) . \tag{2.99}$$

By choosing the step size $\mu^{(i)}$ adequately small with this choice of direction, it is always possible to obtain a reduction of the criterion: $V_N(\theta^{(i+1)}, Z^N) \leq V_N(\theta^{(i)}, Z^N)$. When applying the gradient method to the training of multi-layer perceptron networks it is useful to order the computations in a fashion that utilizes the particular structure of the network. The method is in this case called the *back-propagation algorithm* or the *generalized delta rule*. The back-propagation algorithm will be derived in Section 2.4.3.

The convergence of the method depends alone on the step size $\mu^{(i)}$. Different methods can be applied to select this. A line search generally gives the most rapid convergence in terms of iterations, but it will require a large amount of network evaluations per iteration. Alternatively the step size is frequently controlled adaptively. Examples of adaptive step size control can be found in Hertz et al. (1991) and Demuth and Beale (1998). Also, it is quite often seen that a constant step size is used. As this will have to be selected more or less blindfolded it cannot be recommended.

Regardless of the choice of step size, the (local) convergence rate for the gradient method is linear; that is, there exists a $c \in [0, 1]$ so that $|\theta^{(i+1)} - \theta^*| \leq c|\theta^{(i)} - \theta^*|$. In practice, the gradient method is quite slow. Other methods are available that converge much faster even though their convergence rate, in principle, also is linear. In most applications it is therefore difficult to justify using the gradient method. Nevertheless, the method has gained a remarkable popularity in the neural network community. The primary properties in favor of the method are the simplicity at which it is implemented, and the modest requirement to data storage. An additional property that is sometimes emphasized is the possibility to utilize the parallel structure of the algorithm if the right hardware is available.

In most situations, the drawback associated with the slow convergence motivates the use of more sophisticated methods. Examples of such methods will
be given in the following.

The Newton method. Gradient methods rely on a first-order approximation to the criterion in order to determine the search direction. The Newton
method is the obvious next step. In a Newton method the new iterate is determined as the minimizer of a second-order expansion of the criterion around
the current iterate

$$\tilde{V}_N(\theta, Z^N) = V_N(\theta^{(i)}, Z^N) + \left[\theta - \theta^{(i)}\right]^T G(\theta^{(i)})$$
$$+ \frac{1}{2}\left[\theta - \theta^{(i)}\right]^T H(\theta^{(i)})\left[\theta - \theta^{(i)}\right] . \quad (2.100)$$

Introducing the notation

$$\psi(t, \theta) = \frac{d\hat{y}(t|\theta)}{d\theta} , \quad (2.101)$$

gradient and Hessian for the least squares criterion are given by

$$G(\theta) = V'_N(\theta, Z^N) = \frac{1}{N}\sum_{t=1}^{N}\psi(t, \theta)\left[y(t) - \hat{y}(t|\theta)\right] \quad (2.102)$$

and

$$H(\theta) = V''_N(\theta, Z^N) = \frac{1}{N}\sum_{t=1}^{N}\psi(t, \theta)\psi^T(t, \theta)$$
$$- \frac{1}{N}\sum_{t=1}^{N}\psi'(t, \theta)\varepsilon(t, \theta) . \quad (2.103)$$

The minimum of (2.100) is found at the point where $\tilde{V}'_N(\theta, Z^N) = 0$. Since
the Hessian is symmetric, this corresponds to

$$0 = G(\theta^{(i)}) + \frac{1}{2}\left[2H(\theta^{(i)})\theta - H(\theta^{(i)})\theta^{(i)} - H(\theta^{(i)})\theta^{(i)}\right]$$
$$= G(\theta^{(i)}) + H(\theta^{(i)})\left[\theta - \theta^{(i)}\right] , \quad (2.104)$$

which leads to the following update rule

$$\theta^{(i+1)} = \theta^{(i)} - H^{-1}(\theta^{(i)})G(\theta^{(i)}) . \quad (2.105)$$

This update corresponds to the step size $\mu^{(i)} = 1$, and a search direction
determined by solving the linear system of equations

$$H(\theta^{(i)})f^{(i)} = -G(\theta^{(i)}) . \quad (2.106)$$

The search direction, $f^{(i)}$, is frequently referred to as the Newton direction.

In practice the method must be complemented by a line search because (2.100) is an approximation to the actual criterion. The approximation will usually be valid only in a certain neighborhood around the current iterate and the full step might thus bring the new iterate to a location that is far from the point predicted by the approximation. When a line search is incorporated, the method is sometimes called *the damped Newton method*. In contrast to the gradient method, one cannot be sure that the damped Newton method will converge at all. Except when being close to a minimum, one cannot expect that the criterion is a convex function. If the criterion is not convex, the Hessian will not be positive definite, which implies that (2.100) will be unbounded from below. To circumvent this problem, a gradient method can be used for adjusting the weights in the beginning. When the weights are sufficiently close to the minimum, one can then switch to the damped Newton method. Thereby, the rapid local convergence is retained.

Although the Hessian is positive definite, and the search direction thus is well-defined, it often occurs that the matrix is very ill-conditioned. Consider the series approximation around the minimum, θ^*

$$\tilde{V}_N(\theta, Z^N) = V_N(\theta^*, Z^N) + \frac{1}{2}(\theta - \theta^*)^T H(\theta^*)(\theta - \theta^*) . \qquad (2.107)$$

If a weight θ_j is superfluous then $V_N(\theta^* + \Delta\theta_j, Z^N) = V_N(\theta^*, Z^N)$. Such a weight will thus manifest itself as a singular direction in the Hessian. In the system identification problem, the Hessian should, in principle, never become singular in the minimum. Partly because the system typically will be affected by noise and partly because the condition $S \in M$ will never be satisfied in reality (i.e., the model structure is always insufficient). However, in practice the selected network architecture can be so large that a subset of the weights are not very significant. This causes the Hessian to be nearly singular whereby numerical problems prevents one from computing the search direction. This problem makes the "raw" Newton method improper for training neural networks. A useful remedy is to reduce the condition number for the Hessian by adding a small diagonal matrix to (2.106). This will be further motivated later.

From a computational perspective, calculation of the search direction is an expensive task as it requires the Hessian matrix. Although the convergence rate is quadratic close to a minimum, $|\theta^{(i+1)} - \theta^*| \leq c|\theta^{(i)} - \theta^*|^2$, it happens at the expense of a vary large computational burden in each iteration. Often an approximation to the Hessian that is cheaper to compute is therefore used instead. This class of methods are known as *Quasi-Newton methods*.

By far the most popular scheme for approximating the Hessian is the so-called BFGS algorithm (Broyden-Fletcher-Goldfarb-Shanno). This algorithm produces a positive definite approximation to the Hessian from a consecutive series of previous iterates and corresponding gradients. The algorithm can be implemented in different ways. Often it is done so that the inverse of the Hessian is updated directly

$$\theta^{(i+1)} = \theta^{(i)} + \mu^{(i)} \left[B(\theta^{(i)}) G(\theta^{(i)}) \right] . \tag{2.108}$$

$B(\theta^{(i)}) \simeq H^{-1}(\theta^{(i)})$ is updated according to

$$B^{(i)} = \left[I - \frac{\Delta\theta^{(i)} \left(\Delta G^{(i)}\right)^T}{\left(\Delta G^{(i)}\right)^T \Delta\theta^{(i)}} \right] B^{(i-1)} \left[I - \frac{\Delta G^{(i)} \left(\Delta\theta^{(i)}\right)^T}{\left(\Delta G^{(i)}\right)^T \Delta\theta^{(i)}} \right]$$

$$+ \frac{\Delta\theta^{(i)} \left(\Delta\theta^{(i)}\right)^T}{\left(\Delta G^{(i)}\right)^T \Delta\theta^{(i)}} , \tag{2.109}$$

where the following notation has been applied

$$\Delta\theta^{(i)} \triangleq \theta^{(i)} - \theta^{(i-1)} \qquad \Delta G^{(i)} \triangleq G\left(\theta^{(i)}\right) - G\left(\theta^{(i-1)}\right) \tag{2.110}$$

To ensure that the Hessian approximation remains positive definite, the line search is implemented to comply with the condition

$$\left(\Delta\theta^{(i)}\right)^T \Delta G^{(i)} > 0 . \tag{2.111}$$

See Dennis and Schnabel (1983) for supplementary information.

Although it is often seen that the Quasi-Newton method is proposed for neural network training it rarely pays off to use it directly for nonlinear least squares. Near the minimum the convergence rate (the so-called *local convergence*) is typically good (Dennis and Schnabel, 1983), but the initial convergence tends to be poor. This is not surprising. In the beginning the Hessian matrix will be very different from iteration to iteration, and the approximation of the inverse Hessian will therefore not be very good. Instead it is recommended to consider the family of Gauss-Newton methods, which are derived especially with nonlinear least squares problems in mind.

The Gauss-Newton method. The Gauss-Newton method takes advantage of the following linear approximation to the prediction error, $\varepsilon(t, \theta) = y(t) - \hat{y}(t|\theta)$,

$$\bar{\varepsilon}(t, \theta) = \varepsilon(t, \theta^{(i)}) + \left[\varepsilon'(t, \theta^{(i)})\right]^T (\theta - \theta^{(i)}) \tag{2.112}$$

$$= \varepsilon(t, \theta^{(i)}) - \left[\psi(t, \theta^{(i)})\right]^T (\theta - \theta^{(i)})$$

for modifying the criterion at the ith iteration

$$V_N(\theta, Z^N) \simeq L^{(i)}(\theta) = \frac{1}{2N} \sum_{t=1}^{N} \tilde{\varepsilon}^2(t, \theta) . \qquad (2.113)$$

When evaluated in $\theta = \theta^{(i)}$, the gradient is the same as in the Newton method:

$$G(\theta^{(i)}) = \left. \frac{\mathrm{d}L^{(i)}(\theta)}{\mathrm{d}\theta} \right|_{\theta=\theta^{(i)}} = \frac{1}{N} \sum_{t=1}^{N} \psi(t, \theta^{(i)}) \left[y(t) - \hat{y}(t|\theta^{(i)}) \right] , \qquad (2.114)$$

but the Hessian is different

$$R(\theta^{(i)}) = \left. \frac{\mathrm{d}^2 L^{(i)}(\theta)}{\mathrm{d}\theta^2} \right|_{\theta=\theta^{(i)}} = \frac{1}{N} \sum_{t=1}^{N} \psi(t, \theta^{(i)}) \psi^T(t, \theta^{(i)}) . \qquad (2.115)$$

$R(\theta)$ is called the *Gauss-Newton Hessian*, and it is obviously positive semidefinite by definition. In addition, it has the attractive property that it requires first-order derivative information only. This makes it relatively inexpensive to compute.

Analogous to the Newton method, the Gauss-Newton update is derived as the minimizer of (2.113):

$$\theta^{(i+1)} = \theta^{(i)} - R^{-1}(\theta^{(i)})G(\theta^{(i)}) . \qquad (2.116)$$

In practice, the Gauss-Newton direction is calculated without the inversion but by solving

$$R(\theta^{(i)})f^{(i)} = -G(\theta^{(i)}) . \qquad (2.117)$$

As for the Newton method, the update rule should be complemented with a line search for determining the step size. The method is in this case called the *damped Gauss-Newton method.*

Clearly the Gauss-Newton Hessian equals the true Hessian if the second term of this is neglected

$$H(\theta^{(i)}) = R(\theta^{(i)}) - \frac{1}{N} \sum_{t=1}^{N} \psi'(t, \theta^{(i)})\varepsilon(t, \theta^{(i)}) . \qquad (2.118)$$

This fact is sometimes used for deriving the Gauss-Newton method as a "Poor man's Newton method" rather than obtaining it via the linear approximation of the prediction error as above. It is clear that the Newton and Gauss-Newton methods are identical when the second-order derivative matrix of the network output equals zero

$$\psi'(t, \theta) = \frac{\mathrm{d}^2 \hat{y}(t|\theta)}{\mathrm{d}\theta^2} = 0 . \qquad (2.119)$$

This is the case in linear regression. Apart from this special case, the local convergence of the Gauss-Newton method is only linear. However, for zero or small residual problems, i.e., when the prediction error evaluated in the minimum is zero or small, the second sum of the Hessian will not contribute much close to the minimum. In this situation it seems fair to neglect the second term of the Hessian and apply the Gauss-Newton method. Not surprisingly, it can be shown that the convergence of the Gauss-Newton method is particularly fast on these types of problems (Dennis and Schnabel, 1983). A significant noise level will naturally lead to large residuals, but since the residuals are due to noise they should in the minimum be independent of past data (and in turn ψ). Thus, it is still reasonable to deploy the Gauss-Newton approximation.

Despite the fact that the Gauss-Newton method theoretically provides a slower *local* convergence rate than the Newton (and Quasi-Newton) method, experience shows that it is often faster in practice. In particular when executed far from the minimum.

The Gauss-Newton Hessian will in practice always be positive definite, but it is still subject to the problems with ill-conditioning discussed earlier in relation to the Newton method. It was mentioned that adding a small diagonal matrix to the Hessian before calculating the search direction could alleviate the problem. The *Levenberg-Marquardt method*, which will be presented in a moment, does this in a quite clever fashion.

The Pseudo-Newton method. The so-called Pseudo-Newton method, which is a simplification of the Gauss-Newton method, has gained some popularity in the neural network community (Hertz et al., 1991). The method is based on a crude approximation to the Gauss-Newton Hessian obtained by neglecting all off-diagonal elements. Thus, the weight update is

$$\theta_k^{(i+1)} = \theta_k^{(i)} - \mu^{(i)} G_k(\theta^{(i)}) / R_{kk}(\theta^{(i)}) . \tag{2.120}$$

The advantages of the method are that it avoids solving a large system of equations to obtain the search direction, and that it is less memory-consuming since only the diagonal of the Hessian is needed. Also, the numerical problems associated with the ill-conditioning of the Hessian are overcome. However, the convergence will generally be much slower than for the Gauss-Newton method.

The Levenberg-Marquardt method. The search direction found from a Gauss-Newton (or Newton) method need not be particularly optimal. The direction is determined via an approximation of the criterion, $L^{(i)}(\theta)$, which can be expected to be valid only in a neighborhood around the current iterate. If the minimum of $L^{(i)}(\theta)$ is far from the current iterate, $\theta^{(i)}$, a poor search direction may be obtained. Intuitively, it makes more sense to search for the minimum of $L^{(i)}(\theta)$ only inside some neighborhood around the current iterate. Selecting this neighborhood as a ball with radius $\delta^{(i)}$, the minimization problem can be formulated as

$$\theta^{(i+1)} = \arg\min_\theta L^{(i)}(\theta) \quad \text{subject to} \quad \left|\theta - \theta^{(i)}\right| \leq \delta^{(i)} . \quad (2.121)$$

The update rule appearing when solving the constrained optimization problem can be found in Marquardt (1963):

$$\theta^{(i+1)} = \theta^{(i)} + f^{(i)} \quad (2.122)$$

$$\left[R(\theta^{(i)}) + \lambda^{(i)}I\right] f^{(i)} = -G(\theta^{(i)}) . \quad (2.123)$$

The is a monotonic relationship between the $\lambda^{(i)}$ and the radius, $\delta^{(i)}$, but generally it is not easily found. The method is known as the Levenberg-Marquardt algorithm after the two contributors who independently proposed the algorithm (Levenberg, 1944; Marquardt, 1963). The ball with radius $\delta^{(i)}$ is to be interpreted as the region around $\theta^{(i)}$ inside which the approximation $L^{(i)}(\theta)$ can be trusted to be a valid approximation to the true criterion $V_N(\theta, Z^N)$. The name (model-) trust region method is often used about the method with reference to the this interpretation of the principle (Dennis and Schnabel, 1983; Moré, 1983). The Levenberg-Marquardt method is a trust region method designed specifically for nonlinear least squares. For other optimization problems, the trust region approach must be wrapped around, e.g., a full Newton method.

Unlike the previous methods, it is not common to use line search with the Levenberg-Marquardt method. Basically, it is against the trust region philosophy since the step size in some sense is adjusted automatically along with the trust region radius. To gain some insight into the correspondence between the trust region radius, δ, and the parameter, λ, it is useful to make some heuristic considerations about λ's impact on the search direction. When $R(\theta^{(i)}) + \lambda I$ is substituted for a pure diagonal matrix, the search direction becomes the negative gradient of the criterion. Obviously, when letting $\lambda \to \infty$ the diagonal matrix will dominate over $R(\theta)$, thus leading to the gradient method with a step size approaching zero. On the other hand, if $\lambda = 0$ the Gauss-Newton method is clearly obtained. By adjusting λ, the search direction apparently interpolates between the gradient and the Gauss-Newton direction as illustrated in Figure 2.13.

It is quite obvious that there is close correspondence between reducing the trust region radius and increasing λ (and *vice versa*). Unfortunately, there is generally no closed expression for determining an exact value of λ for a given trust region radius. There are two ways to deal with this: direct methods and indirect methods. The direct methods adjust the trust region directly and then use an iterative method for determination of the corresponding λ. A comprehensive treatment of this approach can be found in Dennis and Schnabel (1983) and Moré (1983). By instead adjusting λ directly, without paying attention to the actual size of the trust region, the indirect methods represent a somewhat less sophisticated approach. The indirect methods

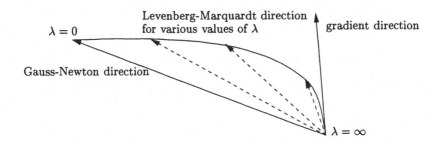

Figure 2.13. The Levenberg-Marquardt direction for various values of λ. As λ is increased, the direction will move from the Gauss-Newton direction ($\lambda = 0$) towards the gradient direction ($\lambda = \infty$) while the step size is reduced to 0.

are similar in spirit to the scheme proposed in the original contribution of Marquardt (1963). To ensure convergence, Marquardt suggested to keep on increasing λ by some factor until a reduction in the criterion occurred. When this was achieved, the iterate was accepted. To avoid that λ would become unnecessarily large it was suggested to decrease it by some factor after each iteration. In Hagan and Menhaj (1994) it is discussed how the method can be applied for neural network training.

Another indirect method is described in Fletcher (1987). This method generally outperforms Marquardt's scheme, but it is equally simple from an implementation perspective. The idea is to observe how well the reduction in the criterion matches the reduction predicted by the approximation $L^{(i)}(\theta)$ and then adjust λ according to this. To measure the accuracy of the approximation, the following ratio is considered

$$r^{(i)} = \frac{V_N(\theta^{(i)}, Z^N) - V_N(\theta^{(i)} + f^{(i)}, Z^N)}{V_N(\theta^{(i)}, Z^N) - L^{(i)}(\theta^{(i)} + f^{(i)})} . \qquad (2.124)$$

If the ratio is close to one, $L^{(i)}(\theta^{(i)} + f)$ is likely to be a reasonable approximation to V_N, and λ should be reduced by some factor (whereby the trust region indirectly is increased). If, on the other hand, the ratio is small or negative, λ should be increased. An algorithm following this principle is outlined in Table 2.1.

The values 0.25, 0.75 and 2 are arbitrary and can be substituted for other values without loss of the convergence properties. Different stopping criteria for use in *Step 6* will be discussed in Section 2.4.3.

Recalling the definition of $L^{(i)}(\theta)$, $L^{(i)}(\theta^{(i)} + f)$ can clearly be written as

$$L^{(i)}(\theta^{(i)} + f) = V_N(\theta^{(i)}, Z^N) + f^T G(\theta^{(i)})$$
$$+ \frac{1}{2} f^T H(\theta^{(i)}) f . \qquad (2.125)$$

Table 2.1. Levenberg-Marquardt algorithm.

---

1 Select an initial parameter vector $\theta^{(0)}$ and an initial value $\lambda^{(0)}$.

2 Determine the search direction from $\left[R(\theta^{(i)}) + \lambda^{(i)}I\right]f^{(i)} = -G(\theta^{(i)})$.

3 $r^{(i)} > 0.75 \Rightarrow \lambda^{(i)} = \lambda^{(i)}/2$.

4 $r^{(i)} < 0.25 \Rightarrow \lambda^{(i)} = 2\lambda^{(i)}$.

5 If $V_N(\theta^{(i)} + f^{(i)}, Z^N) < V_N(\theta^{(i)}, Z^N)$ then accept $\theta^{(i+1)} = \theta^{(i)} + f^{(i)}$ as a new iterate and let $\lambda^{(i+1)} = \lambda^{(i)}$.

6 If the stopping criterion is not satisfied go to *Step 2*.

By re-ordering the expression used for determining the search direction,

$$R(\theta^{(i)})f^{(i)} = -G(\theta^{(i)}) - \lambda f^{(i)} , \qquad (2.126)$$

and inserting it in (2.125), it follows that

$$V_N(\theta^{(i)}, Z^N) - L^{(i)}(\theta^{(i)} + f^{(i)}) = $$
$$\frac{1}{2}\left[-\left(f^{(i)}\right)^T G(\theta^{(i)}) + \lambda^{(i)}\left|f^{(i)}\right|^2\right] , \quad (2.127)$$

which makes the ratio, $r^{(i)}$, inexpensive to compute for use in *Step 3* and *Step 4* of the algorithm.

The quantity $V_N(\theta^{(i)}, Z^N) - L^{(i)}(\theta^{(i)} + f^{(i)})$ is always non-negative. Thus, if the search direction found in *Step 2* does not lead to a reduction in the criterion then $r^{(i)} < 0$, and the inequality in *Step 4* is satisfied. Consequently, λ is always increased until a reduction in the criterion is achieved.

The convergence properties of the Levenberg-Marquardt method are similar to that of the (damped) Gauss-Newton method (Dennis and Schnabel, 1983). In addition, the method has an attractive side effect in the improved numerical conditioning obtained by adding a positive diagonal matrix to the Hessian. Compared to the previously discussed methods, the Levenberg-Marquardt method thus appears to be the most obvious choice for neural network training: a fast and robust convergence is obtained without the need for *ad hoc* solutions. The main drawback of the method is that it is necessary to recompute the search direction every time λ is increased. Even if the weights have not been updated.

A ball is obviously an unsuitable choice of "trust region" when the weights vary greatly in magnitude, and it may have the effect that the convergence is seriously slowed down. Thus, it is sometimes seen that the trust region is defined in a more general fashion by introduction of a scaling matrix, $D^{(i)}$,

$$\left| D^{(i)}(\theta - \theta^{(i)}) \right| \leq \delta^{(i)} . \tag{2.128}$$

The scaling matrix is adjusted simultaneously with the weights in the search for a minimum (Moré, 1983). By using a scaling matrix, the search direction is determined from

$$\left[R(\theta^{(i)}) + \lambda^{(i)} \left(D^{(i)} \right)^T D^{(i)} \right] f^{(i)} = -G(\theta^{(i)}) . \tag{2.129}$$

When neural networks are used in system identification applications, and all signals are scaled to the same variance, experience has shown that weights of very different scales are not a big concern.

Recursive Algorithms. The methods discussed up to now have all been *batch methods*. The expression "batch" refers to the fact that each iteration on the parameter vector requires an evaluation of the entire data set, Z^N. it is sometimes useful to identify a system *on-line*, simultaneously with the acquirement of measurements. Adaptive control is an example of such an application. In this case a model must be identified and a control system designed on-line because the dynamics of the system to be controlled varies with time. Obviously, batch methods are unsuitable in such applications as the amount of computations required in each iteration might exceed the time available within one sampling period. Moreover, old data will be obsolete when the system to be identified is time-varying. The ordinary mean-square error criterion is thus inappropriate for training. In the adaptive control and signal processing communities, several quite sophisticated recursive estimation algorithms have therefore been developed over the years to avoid having to apply batch algorithms on-line.

In a recursive algorithm, one input-output pair from the training set, $[\varphi(t), y(t)]$, is evaluated at a time and used for updating the weights

$$\theta(t) = \theta(t-1) + \mu(t)f(t) . \tag{2.130}$$

Notice that superscript $^{(i)}$ has been substituted by argument t, denoting time (as a multiple of the sampling period).

To cope with time-variations the recursive algorithms usually include a scheme for discarding past information when it is no longer considered valid. Numerous recursive algorithms have been developed of diverse complexity and sophistication in the way past information is discarded (Åström and Wittenmark, 1995; Ljung and Söderström, 1983; Parkum et al., 1992).

It should be emphasized that most recursive algorithms have been developed primarily with identification of linear systems in mind. When the model structure is rich with adjustable parameters, on-line estimation becomes a quite involved business. It is in practice hard to track rapid time-variations, because it requires an amount of new information that measures up to the number of adjustable parameters. Not until data that sufficiently well describes the system under the new operating conditions has been collected, can one expect the weights to be properly adjusted. In addition, different numerical problems will frequently cause problems when there are many adjustable parameters. Because neural networks tend to contain many weights, on-line training is generally dissuaded. Nevertheless, various applications taking advantage of on-line training will be examined in Chapter 3. The emphasis will be on time-invariant systems and the training will be carried out with extreme care.

Although it is recommended here to avoid on-line training algorithms when other solutions are available, recursive algorithms may have a place in off-line training. Instead of assuming that the data set increases with time, which is the case in on-line training, the recursive algorithm is repeated a couple of times on the finite training set, Z^N, collected in advance. This approach may have some advantages over the batch estimation:

- The implementation is simpler.

- Less memory-consuming.

- Redundancy in the data set is utilized more effectively to obtain a faster convergence.

An attempt to depict the difference between batch algorithms, on-line recursive algorithms, and repeated recursive algorithms has been made in Figure 2.14. Recursive off-line schemes for neural network training have been proposed on several occasions, see e.g., Rumelhart and McClelland (1986), Hertz et al. (1991), and Billings et al. (1992).

A recursive Gauss-Newton method. A recursive counterpart to the Gauss-Newton method can be derived quite easily (Ljung, 1999). The derivation is based upon the assumption that new input-output pairs progressively are included in the data set. When used off-line, it is thus assumed that the data set Z^N is repeated the necessary number of times; that is, $u(t) = u(t + N) = u(t + 2N) = \ldots$ and $y(t) = y(t + N) = y(t + 2N) = \ldots$ The criterion considered at time t is then

$$V_t(\theta, Z^t) = \frac{1}{2t} \sum_{k=1}^{t} \varepsilon^2(k, \theta) , \qquad (2.131)$$

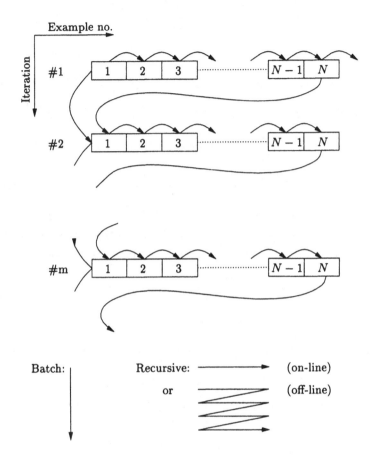

Figure 2.14. The difference between batch algorithms, recursive algorithms for on-line use, and repeated recursive algorithms for off-line use.

and the Gauss-Newton update is consequently given by

$$\theta(t) = \theta(t-1) - R^{-1}(t)V_t'(\theta(t-1), Z^t) . \qquad (2.132)$$

The gradient takes the form

$$V_t'(\theta, Z^t) = -\frac{1}{t}\sum_{k=1}^{t}\psi(k,\theta)\varepsilon(k,\theta)$$

$$= -\frac{1}{t}\sum_{k=1}^{t-1}\psi(k,\theta)\varepsilon(k,\theta) - \frac{1}{t}\psi(t,\theta)\varepsilon(t,\theta)$$

$$= \frac{t-1}{t}V_{t-1}'(\theta, Z^{t-1}) - \frac{1}{t}\psi(t,\theta)\varepsilon(t,\theta) . \qquad (2.133)$$

Assuming that the parameter vector, $\theta(t-1)$, actually minimizes the criterion at time $t-1$, $V_{t-1}(\theta, Z^{t-1})$, implies that the gradient

$$V'_{t-1}(\theta(t-1), Z^{t-1}) = 0 . \qquad (2.134)$$

Thus,

$$V'_{t-1}(\theta, Z^t) = -\frac{1}{t}\psi(t, \theta(t-1))\varepsilon(t, \theta(t-1)) . \qquad (2.135)$$

The expression for the Gauss-Newton Hessian can be rewritten as follows

$$R(t) = \frac{1}{t}\sum_{k=1}^{t}\psi(k, \theta)\psi^T(k, \theta)$$

$$= R(t-1) + \frac{1}{t}\left[\psi(t, \theta)\psi^T(t, \theta) - R(t-1)\right] . \qquad (2.136)$$

By ignoring that $\psi(k, \theta)$, in principle, should be evaluated in $\theta(t-1)$ and not in $\theta(k-1)$, the complete update is given by the following two expressions. $\psi(t) \triangleq \psi(t, \theta(t-1))$ has been introduced to simplify the notation.

$$\theta(t) = \theta(t-1) + \frac{1}{t}R^{-1}\psi(t)\left[y(t) - \hat{y}(t|\theta(t-1))\right] \qquad (2.137)$$

$$R(t) = R(t-1) + \frac{1}{t}\left[\psi(t)\psi^T(t) - R(t-1))\right] . \qquad (2.138)$$

To avoid the inversion of the Gauss-Newton Hessian, the matrix inversion lemma

$$(A + BCD)^{-1} = A - AB(C^{-1} + DAB)^{-1}DA \qquad (2.139)$$

is utilized to derive a formula for direct update of the covariance matrix, $P(t) = \frac{1}{t}R^{-1}(t)$,

$$P(t) = P(t-1) - \frac{P(t-1)\psi(t)\psi^T(t)P(t-1)}{1 + \psi^T(t)P(t-1)\psi(t)} . \qquad (2.140)$$

The user provides the initial covariance matrix $P(0)$. A common choice is $P(0) = cI$ with c being a large number, e.g., $10^4 - 10^8$.

For an ARX model the algorithm degenerates to the so-called *Recursive Least Squares (RLS) algorithm*. In adaptive control it is common to modify the RLS algorithm in various ways to enable a tracking of time-varying systems. Although the recursive Gauss-Newton algorithm in this case primarily is intended to be used off-line, the same type of modifications are recommended to prevent the adaptation from dying out too early. This is necessary because the different assumptions made above are not satisfied in practice and the algorithm therefore must be repeated several times. Some examples of variations of the recursive Gauss-Newton algorithm are given below. See also Åström and Wittenmark (1995), Ljung and Söderström (1983), and Parkum et al. (1992).

Exponential forgetting. A possible strategy for discarding past information is to impose a *forgetting factor* in the criterion:

$$V_t(\theta, Z^t) = \frac{1}{2t} \sum_{k=1}^{t} \lambda^{t-k} \varepsilon^T(k, \theta) \varepsilon(k, \theta) . \tag{2.141}$$

The update can in this case be expressed as follows

$$K(t) = \frac{P(t-1)\psi(t)}{\lambda + \psi^T(t)P(t-1)\psi(t)} \tag{2.142}$$

$$\theta(t) = \theta(t-1) + K(t)\left[y(t) - \hat{y}(t|\theta(t-1))\right] \tag{2.143}$$

$$P(t) = \left[P(t-1) - K(t)\psi^T(t)P(t-1)\right]\frac{1}{\lambda} . \tag{2.144}$$

The forgetting factor λ must be selected in the interval $[0, 1]$, and in practice it is recommended that a value close to unity is used. If λ is too small one is likely to encounter the phenomenon *covariance blow-up*. It occurs when, in certain directions of the parameter space, more past information is forgotten than new information is obtained. This will lead to an increase of certain eigenvalues in the covariance matrix corresponding to an increase in the variance on (and the covariance among) certain parameters. An often-used remedy for covariance blow-up is to impose an upper bound on the eigenvalues in the covariance matrix. The constant trace and EFRA algorithms provide such a property. They additionally impose a lower bound to guarantee that the tracking ability is maintained.

Constant trace.

$$K(t) = P(t-1)\psi(t)\left[1 + \psi(t)^T P(t-1)\psi(t)\right]^{-1} \tag{2.145}$$

$$\theta(t) = \theta(t-1) + K(t)\left[y(t) - \hat{y}(t|\theta(t-1))\right] \tag{2.146}$$

$$\bar{P}(t) = \left[P(t-1) - \frac{P(t-1)\psi(t)\psi^T(t)P(t-1)}{1 + \psi^T(t)P(t-1)\psi(t)}\right]\frac{1}{\lambda} \tag{2.147}$$

$$P(t) = \frac{\alpha_{max} - \alpha_{min}}{\text{tr}\left\{\bar{P}(t)\right\}}\bar{P}(t) - \alpha_{min}I . \tag{2.148}$$

α_{min} and α_{max} are the minimum and maximum eigenvalues, respectively. They are typically selected so that $\alpha_{max}/\alpha_{min} \simeq 10^4$. The initial covariance matrix should be selected as a diagonal matrix, $\alpha_{min}I < P(0) < \alpha_{max}I$.

Exponential Forgetting and Resetting Algorithm (EFRA).

$$K(t) = \alpha P(t-1)\psi(t)\left[1 + \psi^T(t)P(t-1)\psi(t)\right]^{-1} \tag{2.149}$$

$$\theta(t) = \theta(t-1) + K(t)\left[y(t) - \hat{y}(t|\theta(t-1))\right] \tag{2.150}$$

$$P(t) = \frac{1}{\lambda}P(t-1) - K(t)\psi^T(t)P(t-1) + \beta I - \delta P^2(t-1) . \tag{2.151}$$

The EFRA algorithm was proposed in Salgado et al. (1988). The four design parameters α, β, δ and λ must be selected so that they satisfy

$$0 < \gamma < \alpha < 1$$
$$(\gamma - \alpha)^2 + 4\beta\delta < (1 - \alpha)^2 \qquad (2.152)$$
$$0 < \beta, \quad 0 < \delta,$$

where $\gamma \triangleq \frac{1-\lambda}{\lambda}$. The minimum and maximum eigenvalues are determined as follows

$$\alpha_{min} = \left(\frac{\alpha - \gamma}{2\delta}\right)\left(\sqrt{1 + \frac{4\beta\delta}{(\alpha - \gamma)^2}} - 1\right) \qquad (2.153)$$

$$\alpha_{min} = \frac{\gamma}{2\delta}\left(\sqrt{1 + \frac{4\beta\delta}{\gamma^2}} + 1\right). \qquad (2.154)$$

A recursive gradient method. A recursive version of the gradient method is obtained by setting $\frac{1}{t}R^{-1}(t) = \mu I$. In the neural network community this is frequently referred to as *incremental* or *on-line* back-propagation (Hertz et al., 1991).

2.4.2 Regularization and the Concept of Generalization

Up to now it has been encouraged to apply the training set, Z^N, for picking a model from a model structure according to a mean square error criterion. Before continuing the journey through the identification procedure, it is relevant to reflect on whether the mean square error at all is a sensible criterion to work with, or it is possible to find something better. Based on fairly heuristic statistical considerations, the present section attempts to shed some light on the properties of neural network models trained according to a mean square error criterion as well as other criteria. More in-depth treatments of the topic can be found in Ljung (1999), Geman et al. (1992), and Sjöberg and Ljung (1995).

Assume that the system under consideration is governed by some function of past data plus white noise independent of the input

$$y(t) = f(Z^{t-1}) + e(t). \qquad (2.155)$$

A neural network with a finite number of weights is in general able to model the true system only to a certain accuracy. Nevertheless, as mentioned before it is sometimes useful to think of the data set as if it was actually generated by a "true" neural network model,

$$y(t) = g_0[\varphi(t, \theta_0), \theta_0] + e(t). \qquad (2.156)$$

Although this can never be an exact description, it is usually fair to neglect the inaccuracy if only the network is reasonably large.

To recapitulate the background for the least squares training explored in Section 2.4.1, the principle is that Z^N is used for picking a model from the model structure $\mathcal{M}$, which has p weights,

$$\hat{y}(t|\theta) = g[\varphi(t,\theta),\theta] , \qquad \theta \in \mathcal{D}_m \subset \mathbb{R}^p \qquad (2.157)$$

according to

$$\hat{\theta} = \arg\min_\theta V_N(\theta, Z^N) = \arg\min_\theta \frac{1}{2N} \sum_{t=1}^{N} [y(t) - \hat{y}(t|\theta)]^2 . \qquad (2.158)$$

A better measure than V_N would be the expectation of the prediction error with respect to the regression vector, $\varphi(t)$, and the noise contribution, $e(t)$. This quantity is called the *generalization error*:

$$\bar{V}(\theta) = \frac{1}{2}\mathbf{E}\left\{[y(t) - \hat{y}(t|\theta)]^2\right\} . \qquad (2.159)$$

Unfortunately it is not possible to evaluate this criterion in practice, but under suitable conditions (Ljung, 1999)

$$\lim_{N\to\infty} V_N(\theta, Z^N) = \bar{V}(\theta) \qquad (2.160)$$

and thus $\hat{\theta} \to \theta^*$ for $N \to \infty$, where θ^* is the minimizer of the generalization error. In particular, if the true system described by (2.156) was actually contained in the model structure, $\mathcal{S} \in \mathcal{M}$, the estimate would also be consistent: $\theta^* = \theta_0$. In reality the training set will be finite, and it is more relevant to consider what to expect of the trained network in this situation. In Ljung (1999) the discrepancy between $\hat{\theta}$ and θ^* that is due to a finite training set is investigated. It is shown that the weights converge in distribution to a Gaussian distribution (the asymptotic Gaussian is denoted AsN(mean, covariance matrix)),

$$\hat{\theta} \in AsN(\theta^*, \frac{1}{N}P_\theta) . \qquad (2.161)$$

If $\mathcal{S} \in \mathcal{M}$, the asymptotical covariance matrix is given by

$$P_\theta = \sigma_e^2 \left[\mathbf{E}\left\{\psi(t,\theta_0)\psi^T(t,\theta_0)\right\}\right]^{-1}$$

$$\simeq 2V_N(\hat{\theta}, Z^N) \left[\frac{1}{N}\sum_{t=1}^{N}\psi(t,\hat{\theta})\psi^T(t,\hat{\theta})\right]^{-1} . \qquad (2.162)$$

If $\mathcal{S} \notin \mathcal{M}$, it is in general more difficult to obtain a useful expression for P_θ.

An estimate of the generalization error is commonly obtained by evaluating the trained neural network model on a *test* (or *validation*) set; a data set, Z^T, containing data that was not used for training. If $\bar{V}(\hat{\theta}) \simeq V_T(\hat{\theta}, Z^T)$ is close to $V_N(\hat{\theta}, Z^N)$ it is likely that $\hat{\theta}$ is close to θ^* and that the trained model therefore is reasonably good.

If it is possible to set aside only a small data set for testing (or none at all), it is difficult to produce a reliable estimate of the generalization error. Another limitation of the generalization error is that it indirectly depends on Z^N through $\hat{\theta}$. That is, it does not measure how well the model structure has been selected, i.e., how one can *expect* a model trained on a data set of size N to perform on unseen data. For this reason it is common to introduce the *average generalization error* as a measure of the model quality

$$V_M = \mathbf{E}\left\{\bar{V}(\hat{\theta})\right\} . \tag{2.163}$$

The expectation here is with respect to training sets of size N. It is possible to obtain fairly simple estimates of this quantity. One example is Akaike's Final Prediction Error (FPE) estimate (see for example Ljung (1999)), which is valid for $\mathcal{S} \in \mathcal{M}$

$$\hat{V}_M = \frac{1}{2}\sigma_e^2 \left(1 + \frac{p}{N}\right) . \tag{2.164}$$

The best possible generalization error is one half the noise variance, $\frac{1}{2}\sigma_e^2$, but because the training set is finite the average generalization error will always be somewhat greater. When analyzing the reasons why the average generalization error for a network exceeds $\frac{1}{2}\sigma_e^2$, it is useful to separate the modelling error into two contributions:

The bias error. The portions of the error that are due to an insufficient model structure; i.e., $\mathcal{S} \in \mathcal{M}$ is not satisfied, and an insufficient sample size. The bias error is defined as the error of the function obtained by averaging over all data sets of the given size N: $f[\varphi(t)] \triangleq \mathbf{E}\left\{g[\varphi(t), \hat{\theta}]\right\}$

The variance error. The portion of the error that is due to the fact that the function implemented by the network obtained on a specific data set, $g[\varphi(t), \hat{\theta}]$, deviates from the average function, $f[\varphi(t))]$.

The decomposition of the error in bias and variance components is straightforward

$$V_M = \mathbf{E}\left\{\bar{V}(\hat{\theta})\right\}$$

$$= \mathbf{E}\left\{\left(g_0[\varphi(t), \theta_0] - g[\varphi(t), \hat{\theta}]\right)^2\right\} + \sigma_e^2$$

$$= \mathbf{E}\left\{\left(g_0[\varphi(t), \theta_0] - \mathbf{E}\left\{g[\varphi(t), \hat{\theta}]\right\}\right)^2\right\}$$

$$+ \quad \mathbf{E}\left\{\left(\mathbf{E}\left\{g[\varphi(t), \hat{\theta}]\right\} - g[\varphi(t), \hat{\theta}]\right)^2\right\} + \sigma_e^2 . \qquad (2.165)$$

As mentioned above, $S \notin M$ is in practice inevitable and thus there will always be a certain, maybe small, bias error; even in the limit of large training data sets. However, typically one will find that the bias error is decreased as more weights are added since the network will be able to describe the system more accurately. The reason for not just selecting a gigantic network architecture is that one must expect the variance error to work in the opposite direction: as more weights are estimated on the same data set, the variance on the estimated weights will increase. This quandary is often referred to as the *bias/variance dilemma*, and it is a dilemma which frequently appears in different disguises. For a discussion of the dilemma in a neural network context one may consult Geman et al. (1992) and Heskes (1998).

To demonstrate the dilemma in practice, consider the following experiment:

Train ten different NNARX(2,2,1) models on the same set of noisy measurements, Z^N, where $N = 400$. Let the simplest model structure have one hidden unit and increase the model structure gradually with one hidden unit at a time (corresponding to six weights). When the networks have been trained to a minimum they are evaluated on an additional data set, Z^T, consisting of 2000 samples. The quantity, $V_T(\hat{\theta}, Z^T)$, is interpreted as an estimate of the (average) generalization error. Due to the problems with multiple minima, each network is trained five times using different initializations. The result of the experiment is shown in Figure 2.15. It appears from Figure 2.15 that the optimal trade-off between bias and variance error is achieved for a network containing four or five hidden units. If the network has more hidden units, the variance error dominates. To describe this situation, it is common to use the expression *overfitting*. Overfitting means that the network not only models the features of the system, but to an undesired extent also the noise in the training set. Likewise, the expression *underfitting* is used when the bias error is dominating.

If one is interested in the best possible accuracy, model structure selection is much more involved than just a matter of selecting a number of hidden units. The network need not be fully connected; in fact, it is likely that it will be advantageous to leave out weights connecting certain inputs with certain hidden units or certain hidden units with certain outputs. Obviously

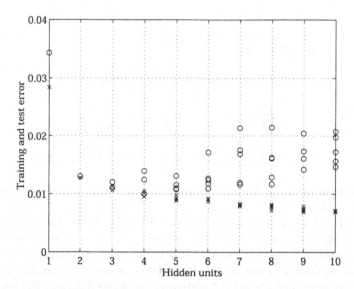

Figure 2.15. Ten different NNARX models of i different size trained five times on the same set of data. 'x' marks the training error, $V_N(\hat{\theta}, Z^N)$, while 'o' marks the test error, $V_T(\hat{\theta}, Z^T)$.

it will lead to a combinatorial explosion if one was to pick the model on a trail-and-error basis as in the example above. In Section 2.6 some methods for finding the optimal network architecture automatically will be discussed. Another way to deal with the bias/variance dilemma is to augment the criterion $V_N(\theta, Z^N)$ with a *regularization* (or complexity) term. The most commonly used augmentation is the so-called *simple weight decay* term

$$W_N(\theta, Z^N) = \frac{1}{2N} \sum_{t=1}^{N} [y(t) - \hat{y}(t|\theta)]^2 + \frac{1}{2N} \theta^T D\theta . \qquad (2.166)$$

D is a diagonal matrix and is most often selected as $D = \alpha I$, where α is denotes the *weight decay*. Sometimes a different weight decay is used for the input-to-hidden layer and hidden-to-output layer weights, respectively, or each weight is affiliated with an individual weight decay parameter. Note that the influence of the weight decay in general will decrease as $N \to \infty$, hence the bias will vanish for large data sets. Regularization imposes a bias towards zero, hence, will increase the bias error for finite data sets. What about the variance error then? In Sjöberg and Ljung (1995) the FPE estimate has been derived for networks trained according to (2.166) with $D = \alpha I$ (again under the assumption $S \in \mathcal{M}$)

$$\hat{V}_M = \frac{1}{2} \left[\sigma_e^2 \left(1 + \frac{p_1}{N} \right) + \gamma \right] , \qquad (2.167)$$

where

$$p_1 = \text{tr}\left\{ R \left[R + \frac{\alpha}{N}I \right]^{-1} R \left[R + \frac{\alpha}{N}I \right]^{-1} \right\} \tag{2.168}$$

$$\gamma = \frac{\alpha^2}{N^2}\, \theta_0^T \left[R + \frac{\alpha}{N}I \right]^{-1} R \left[R + \frac{\alpha}{N}I \right]^{-1} \theta_0 \; \leq \; \frac{\alpha}{4N}|\theta|^2 \tag{2.169}$$

and

$$R = \mathbf{E}\left\{ \psi(t,\theta_0)\psi^T(t,\theta_0) \right\} \simeq \frac{1}{N}\sum_{t=1}^{N} \psi(t,\hat{\theta})\psi^T(t,\hat{\theta}) . \tag{2.170}$$

Since the trace of a matrix equals the sum of its eigenvalues, p_1 can be written as

$$p_1 = \sum_{i=1}^{p} \frac{\delta_i}{\left(\delta_i + \frac{\alpha}{N}\right)^2} , \tag{2.171}$$

where δ_i specifies the ith eigenvalue of the Hessian matrix, R. The modifications of the FPE estimate for regularized models were also analyzed in Moody (1991).

As discussed in Section 2.4.1, each superfluous weight will lead to an eigenvalue in the Hessian that is zero. In practice no weight will be *completely* superfluous, because the network always represents an underparametrization. The Hessian is therefore always positive definite. However, a weight of little importance will in general lead to a small eigenvalue and *vice versa*. This can also be seen by recognizing the derivative of the output with respect to the weights, $\psi(t,\theta)$, as a "sensitivity" matrix. If weight i is insignificant, the derivative of this particular weight will be small for all t. This has the effect that the diagonal element as well as all elements in row i (and column i) in the matrix R will be small, resulting in a small eigenvalue. For more significant weights the opposite can be expected. One can therefore split the eigenvalues into two separate groups: a group of small eigenvalues that are due to weights of little importance and a group of large eigenvalues that are due to more significant weights. If it is assumed that α/N is much bigger than the smallest eigenvalues and much smaller than the biggest eigenvalues, the quantity p_1 counts the number of significant weights. Hence, p_1 is referred to as the *effective number of weights* in the network. With this interpretation, and by ignoring γ, the new FPE estimate is indeed similar in structure to the FPE estimate for an unregularized criterion. Using the weight decay as a tuning knob one can in essence determine the effective size of the network architecture. The challenge is then to select the weight decay such that the average generalization error is minimized.

It should be mentioned that neither of the FPE estimates (2.164) and (2.167) can be calculated directly as the expressions are partly unknown (the noise variance is not available). In Section 2.5 it will be discussed how to estimate the noise variance.

An effect similar to regularization can be accomplished by stopping the training session before the minimum of the criterion has been reached. This is easy to illustrate in practice. Consider a heavily overparametrized NNARX model structure with 15 hidden units that is trained and evaluated on the data sets from the previous experiment. Train the network with the Levenberg-Marquardt method and evaluate the test error after each iteration.

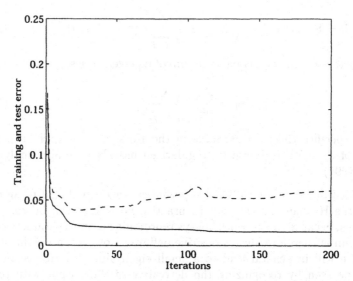

Figure 2.16. A typical example of the development of the criterion (=the training error), $V_N(\theta^{(i)}, Z^N)$, and the test error, $V_T(\theta^{(i)}, Z^T)$, during a training session. The network is trained with the Levenberg-Marquardt method. The training error is represented by the solid line and the test error by the dashed line.

Due to the implementation of the Levenberg-Marquardt algorithm, the training error is of course a monotonic decreasing function of the iteration. Notice that the test error is only reduced in the beginning. After about 25 iterations it reaches a minimum and then starts growing. Loosely speaking, one can explain this as the network initially captures the essential features of the system but after a while it adapts mostly to the noise. In other words, after 25 iterations the network starts overfitting. Hence, this phenomenon is referred to as *overtraining*.

In Sjöberg and Ljung (1995) the interesting result was shown that the effect of early stopping not only has an effect similar to regularization, but that the two approaches are in fact closely related. However, it is usually preferred to use regularization over early stopping. In Section 2.5 and Section 2.6 it will be shown that the derivation of various important quantities related to validation and optimal architecture selection depends on the network being in a minimum. This condition is not satisfied with early stopping.

The training session depicted in Figure 2.16 was performed according to the unregularized criterion. To get a grasp of the difference it is interesting to repeat the experiment with the regularized criterion. The result is shown in Figure 2.17. In the following section, Section 2.4.3, it is explained how the training methods should be modified to cope with the weight decay term.

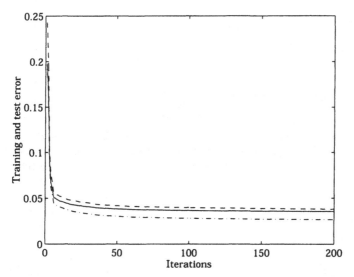

Figure 2.17. Training session with the regularized criterion $W_N(\theta^{(i)}, Z^N)$. $W_N(\theta^{(i)}, Z^N)$ is shown as a solid line, $V_T(\theta^{(i)}, Z^T)$ is the dashed line, and $V_N(\theta^{(i)}, Z^N)$ is the dot-dashed line. The weight decay is $\alpha = 1.0$.

It is clear from Figure 2.17 that the weight decay eliminates the overtraining phenomenon and that the final value of the test error matches the result obtained by early stopping. Moreover, the minimum of the criterion appears to have been reached in less iterations than for the unregularized case.

Naturally, it is not unimportant how the weight decay is selected. If, for instance, $\alpha = 100$ the less attractive result shown in Figure 2.18 is obtained.

The most reliable method for selecting the optimal weight decay is typically through a trial-and-error experiment with a representative test set. The result of such an experiment is shown in Figure 2.19 for seventy different choices of weight decay in the interval $10^{-5} < \alpha < 10^3$. To account for the occurrence of local minima, the training algorithm was for each choice of weight decay executed five times starting with different initial weights.

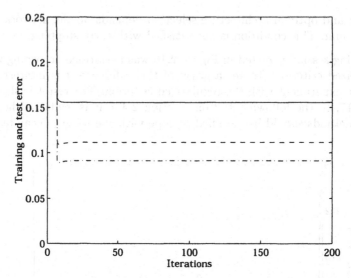

Figure 2.18. Training session according with the regularized criterion $W_N(\theta^{(i)}, Z^N)$. $W_N(\theta^{(i)}, Z^N)$ is shown as a solid line, $V_T(\theta^{(i)}, Z^T)$ is the dashed line and $V_N(\theta^{(i)}, Z^N)$ is the dot-dashed line. The weight decay is $\alpha = 100$.

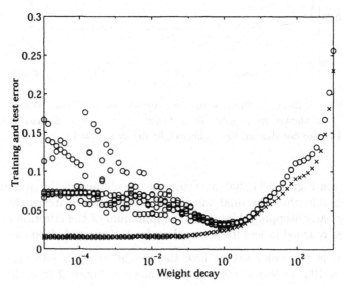

Figure 2.19. A network with 15 hidden units trained with 70 different weight decays. For each choice of weight decay, the network is trained 5 times using different initializations. The final value of the training error, $V_N(\hat{\theta}, Z^N)$, is shown with an 'x' while the test error, $V_T(\hat{\theta}, Z^T)$, is shown with an 'o'.

Figure 2.19 shows that weight decay has other interesting features than just to improve generalization. With a small (or no) weight decay there is not only a large gap between training and test error, there is also a large spread in the test error corresponding to each individual local minimum. As the weight decay is increased, the gap between training and test error is decreased and so is the spread in the errors of the different minima. In fact, it seems as though the local minima gradually are being eliminated as the weight decay is increased. This feature is not surprising. The regularization term represents a simple function of the weights, and it has a smoothing effect on the criterion (Pedersen and Hansen, 1995). The increased smoothness also explains why the minimum of the criterion is reached faster when the network is trained with weight decay (compare Figure 2.16, Figure 2.17, and Figure 2.18). A smoother criterion implies that the approximation made by the Levenberg-Marquardt method in each iteration will be more accurate and consequently it is possible to take bigger steps.

Unfortunately, the trial-and-error test depicted in Figure 2.19 will be too demanding for large network architectures and large training sets. In particular if a more advanced regularization scheme with multiple weight decays is desired. Different attempts have been made to select the weight decay simultaneously with the training of the network. See, e.g., Hansen et al. (1994), MacKay (1992a), MacKay (1992b), and Larsen et al. (1998).

2.4.3 Remarks on Implementation

Below a few hints are given regarding the implementation aspects of neural network training methods.

Training with weight decay. The training methods presented in Section 2.4.1 apply more or less directly to the regularized criterion, $W_N(\theta^{(i)}, Z^N)$. It is necessary to add additional terms to the gradient and the Hessian

$$G(\theta) = W_N'(\theta, Z^N) = V_N'(\theta, Z^N) + \frac{1}{N} D\theta \tag{2.172}$$

$$H(\theta) = W_N''(\theta, Z^N) = V_N''(\theta, Z^N) + \frac{1}{N} D . \tag{2.173}$$

The Levenberg-Marquardt method requires some extra modifications to conform to the regularized criterion. The Gauss-Newton approximation to $W_N(\theta^{(i)}, Z^N)$ is

$$W_N(\theta, Z^N) \simeq L^{(i)}(\theta) = \frac{1}{2N} \left(\sum_{t=1}^{N} [\bar{\varepsilon}(t, \theta)]^2 + \theta^T D\theta \right) , \tag{2.174}$$

which has the Hessian

$$R(\theta) = \left.\frac{\mathrm{d}^2 L^{(i)}(\theta)}{\mathrm{d}\theta^2}\right|_{\theta=\theta^{(i)}} = \frac{1}{N}\left(\sum_{t=1}^{N}\psi(t,\theta^{(i)})\psi^T(t,\theta^{(i)}) + D\right). \quad (2.175)$$

The ratio used for adjusting the Levenberg-Marquardt parameter, λ, must obviously be changed to

$$r^{(i)} = \frac{W_N(\theta^{(i)}, Z^N) - W_N(\theta^{(i)} + f^{(i)}, Z^N)}{W_N(\theta^{(i)}, Z^N) - L^{(i)}(\theta^{(i)} + f^{(i)})}. \quad (2.176)$$

As before the denominator, $W_N(\theta^{(i)}, Z^N) - L^{(i)}(\theta^{(i)} + f^{(i)})$, can be calculated by quite simple means. Straightforward matrix manipulations give

$$W_N(\theta^{(i)}, Z^N) - L^{(i)}(\theta^{(i)} + f^{(i)}) =$$
$$\frac{1}{2}\left[\left(f^{(i)}\right)^T\left(G(\theta^{(i)}) + \left[\lambda^{(i)}I + \frac{1}{N}D\right]f^{(i)}\right)\right]. \quad (2.177)$$

It is worth noting that weight decay improves the numerical robustness of the second-order training algorithms (i.e., the algorithm using the Hessian or an estimate of it). As the expressions for calculation of Hessians (2.173) and (2.175) reveal, the condition number is directly reduced by the addition of positive factors in the diagonal. In fact, this will in many cases enable straightforward implementation of the (damped) Newton and Gauss-Newton methods.

Computing the gradients. Except for the full Newton method, which also requires second-order derivative information, the derivative of the prediction with respect to the weights, $\psi(t,\theta)$, is the only component that really impedes the implementation of the training methods. This subsection provides formulas for calculating the derivatives for the model structures presented previously.

$\psi(t,\theta)$ *for NNARX and NNFIR models:*

$$\psi(t,\theta) = \frac{\mathrm{d}\hat{y}(t|\theta)}{\mathrm{d}\theta} = \frac{\partial\hat{y}(t|\theta)}{\partial\theta} = \phi(t). \quad (2.178)$$

$\psi(t,\theta)$ *for NNARMAX models:*

$$\psi(t,\theta) = \frac{\mathrm{d}\hat{y}(t|\theta)}{\mathrm{d}\theta} = \frac{\partial\hat{y}(t|\theta)}{\partial\theta} - \frac{\partial\hat{y}(t|\theta)}{\partial\varepsilon(t-1,\theta)}\frac{\partial\hat{y}(t-1|\theta)}{\partial\theta} -$$
$$\cdots - \frac{\partial\hat{y}(t|\theta)}{\partial\varepsilon(t-k,\theta)}\frac{\partial\hat{y}(t-k|\theta)}{\partial\theta}$$
$$= \phi(t) - c_1(t)\psi(t-1,\theta) - \cdots - c_k(t)\psi(t-k,\theta) \quad (2.179)$$

or, by introduction of $C(t,q^{-1}) = 1 + c_1(t)q^{-1} + \cdots + c_k(t)q^{-k}$,

$$\psi(t,\theta) = \frac{1}{C(t,q^{-1})}\phi(t) \ . \tag{2.180}$$

The price paid for letting the regression vector depend on the weights in the network is obvious when comparing (2.180) to (2.178): for NNARMAX models the gradient is calculated by filtering the partial derivative $\phi(t)$ with a time-varying linear filter. It is evident that it will cause stability problems if the filter denominator, $C(t,q^{-1})$, frequently has roots outside the unit circle. However, as the training algorithm approaches the minimum, this should not occur except perhaps for a few data points. Unfortunately, in the first few iterations one may encounter numerical problems due to instability if the weights are just initialized at random. If the NNARMAX model is trained with a recursive algorithm, the problems with instability of the filter often become quite serious because there is no line search involved with the recursive algorithms. Thus, it might be necessary to execute the algorithm more than once to obtain a solution.

$\psi(t,\theta)$ *for NNOE models:*

$$\begin{aligned}
\psi(t,\theta) = \frac{d\hat{y}(t|\theta)}{d\theta} &= \frac{\partial\hat{y}(t|\theta)}{\partial\theta} + \frac{\partial\hat{y}(t|\theta)}{\partial\hat{y}(t-1,\theta)}\frac{\partial\hat{y}(t-1|\theta)}{\partial\theta} + \\
&\quad \cdots + \frac{\partial\hat{y}(t|\theta)}{\partial\hat{y}(t-r,\theta)}\frac{\partial\hat{y}(t-r|\theta)}{\partial\theta} \\
&= \phi(t) - f_1(t)\psi(t-1,\theta) - \cdots - f_r(t)\psi(t-r,\theta) \tag{2.181}
\end{aligned}$$

or, with $F(t,q^{-1}) = 1 + f_1(t)q^{-1} + \cdots + f_r(t)q^{-r}$,

$$\psi(t,\theta) = \frac{1}{F(t,q^{-1})}\phi(t) \ . \tag{2.182}$$

The output error model is subject to the same type of stability problems as the NNARMAX model. If the system to be identified is stable (and otherwise one should not consider an output error model), the polynomial, F, will in general have its roots inside the unit circle near the minimum.

$\psi(t,\theta)$ *for NNSSIF models:*

Computation of the gradient for a model on state space innovations form is slightly more involved than for the input-output models considered above. By assuming that all the states can be measured, the gradient is given by

$$\psi(t,\theta) = \frac{d\hat{x}^T(t|\theta)}{d\theta}C^T = \psi_x(t,\theta)C^T \tag{2.183}$$

where

$$\psi_x(t,\theta) = \frac{\mathrm{d}\hat{x}^T(t|\theta)}{\mathrm{d}\theta} = \frac{\partial \hat{x}^T(t|\theta)}{\partial \theta} + \frac{\mathrm{d}\hat{x}^T(t-1|\theta)}{\mathrm{d}\theta} \frac{\partial \hat{x}^T(t|\theta)}{\partial \hat{x}^T(t-1|\theta)}$$

$$- \frac{\mathrm{d}\hat{y}^T(t-1|\theta)}{\mathrm{d}\theta} \frac{\partial \hat{x}^T(t|\theta)}{\partial \varepsilon(t-1,\theta)}$$

$$= \phi(t) + \psi_x(t-1,\theta)A^T(t) - \psi(t-1,\theta)K^T(t)$$

$$= \phi(t) + \psi_x(t-1,\theta)\left[A(t) - K(t)C\right]^T . \tag{2.184}$$

If one does not posses full state information it is necessary to separate the state prediction into two networks as discussed in Section 2.2.2. However, when computing the gradient it is perhaps simpler to neglect this by treating the model as a single network. In this case, one has to enforce in the matrix $A(t)$ that for all $j \in \{q_i\}$ and all $k \in \{q_i + 1\}$

$$A_{j,k} = \begin{cases} 1, j = k+1 \\ 0, \text{otherwise} \end{cases} . \tag{2.185}$$

The calculation of the remaining entries in $A(t)$ and the matrix $K(t)$ is shown later (see (2.189)).

The partial derivative, $\phi(t)$:

$\phi(t)$ is the derivative of the prediction (the network output) with respect to the weights when the regressor's dependency on the weights is ignored. Let $\hat{z}_k$ be a generic output variable symbolizing either $\hat{y}$ or $\hat{x}_k$. For a two-layer network with *tanh* hidden units and linear output units,

$$\hat{z}_k(t|\theta) = g_k[\varphi(t,\theta),\theta] = \sum_{j=1}^{n_h} W_{k,j} \tanh \left(\sum_{l=1}^{n_\varphi} w_{j,l}\varphi_l(t,\theta) + w_{j,0} \right) + W_{k,0}$$

$$= \sum_{j=1}^{n_h} W_{k,j} h_j(t,\theta) + W_{k,0} , \tag{2.186}$$

the partial derivatives are calculated as follows

$$\frac{\partial \hat{z}_k(t|\theta)}{\partial W_{i,j}} = \begin{cases} h_j(t), j > 0, k = i \\ 1, \qquad j = 0, k = i \\ 0, \qquad i \neq k \end{cases} \tag{2.187}$$

$$\frac{\partial \hat{z}_k(t|\theta)}{\partial w_{j,l}} = \begin{cases} W_{k,j}\left[1 - h_j^2(t)\right]\varphi_l(t,\theta), l > 0 \\ W_{k,j}\left[1 - h_j^2(t)\right], \qquad l = 0 \end{cases} \tag{2.188}$$

$\phi(t)$ is a matrix with as many rows as there are weights in the network and as many columns as there are outputs. How the elements specified in (2.187) and (2.188) are ordered in the $\phi(t)$-matrix is of course purely a matter of choice.

Instantaneous Jacobian of a neural network model:

The *instantaneous Jacobian* or *actual gain matrix* (Sørensen, 1996, 1993) is the derivative of the network output with respect to the inputs for a given input-output pair. The Jacobian is required for a portion of the inputs when the regression vector depends on the weights. Considering a two-layer network with *tanh* hidden units and linear output units, the partial derivative of a network output with respect to an arbitrary input is calculated according to

$$\frac{\partial \hat{z}_k(t|\theta)}{\partial \varphi_l(t,\theta)} = \sum_{j=1}^{n_h} W_{k,j} w_{j,l} \left[1 - \tanh^2 \left(\sum_{l=1}^{n_\varphi} w_{j,l} \varphi_l(t,\theta) + w_{j,0} \right) \right]$$

$$= \sum_{j=1}^{n_h} W_{k,j} w_{j,l} \left[1 - h_j^2(t,\theta) \right] . \qquad (2.189)$$

Back-propagation:

When networks with more than one layer of nonlinear activation functions are considered, the expressions for the elements in the partial derivative matrix, $\phi(t)$, obviously become more complex. In particular, one should be careful not to carry out the same computations more than once. Due to the simple structure it is possible to derive a very nice algorithm for computing the gradient of the criterion for general n-layer feedforward networks with arbitrary activation functions. The algorithm is typically referred to as *(error) back-propagation* or the *generalized delta rule* (Hertz et al., 1991). The algorithm is shown below for a two-layer network. Since the basic algorithm is restricted to pure feedforward networks, it can only be used with NNFIR and NNARX model structures. However, the algorithm is quite easily modified to give only the partial derivative matrix, $\phi(t)$.

It is recalled that the gradient of the least squares criterion takes the form (for completeness the multi-output case is now considered)

$$G(\theta^{(i)}) = V_N'(\theta^{(i)}, Z^N) = \frac{1}{N} \sum_{t=1}^{N} \frac{\partial \varepsilon^T(t,\theta)}{\partial \theta} \varepsilon(t,\theta) \bigg|_{\theta=\theta^{(i)}}$$

$$= -\frac{1}{N} \sum_{t=1}^{N} \frac{\partial \hat{y}^T(t|\theta)}{\partial \theta} \left[y(t) - \hat{y}(t|\theta) \right] \bigg|_{\theta=\theta^{(i)}} \qquad (2.190)$$

The output of the kth unit of a general two-layer network can be expressed as

$$\hat{y}(t|\theta) = F_k \left[\sum_{j=0}^{n_h} W_{k,j} h_j(t) \right]$$

$$= F_k \left[\sum_{j=1}^{n_h} W_{k,j} f_j \left(\sum_{l=0}^{n_\varphi} w_{j,l} \varphi_l(t) \right) + W_{k,0} \right] . \qquad (2.191)$$

f_j specifies the activation function for unit j in the hidden layer and F_k specifies the activation function for output k. For simplicity, the biases are regarded as additional weights; i.e., $h_0(t) = \varphi_0(t) = 1$.

The partial derivatives of the network output with respect to the weights in each of the two layers are determined by

$$\phi_{j,k}^{(W)} = \frac{\partial \hat{y}_k(t|\theta)}{\partial W_{k,j}} = h_j(t) F_k' \left[\sum_{j=0}^{n_h} W_{k,j} h_j(t) \right] \tag{2.192}$$

and

$$\phi_{j,k,l}^{(w)} = \frac{\partial \hat{y}_k(t|\theta)}{\partial w_{k,l}}$$

$$= \varphi_l(t) f_j' \left[\sum_{j=0}^{n_\varphi} w_{j,l} \varphi_l(t) \right] W_{k,j} F_k' \left[\sum_{j=0}^{n_h} W_{k,j} h_j(t) \right] . \tag{2.193}$$

This leads to the following expression for the gradient in the hidden-to-output layer

$$G(W_{k,j}) = \sum_{t=1}^{N} h_j(t) F_k' \left[\sum_{j=0}^{n_h} W_{k,j} h_j(t) \right] (y_k(t) - \hat{y}_k(t|\theta))$$

$$= \sum_{t=1}^{N} h_j(t) \delta_k^{(W)}(t) , \tag{2.194}$$

where the quantity $\delta_k^{(W)}(t)$ has been introduced

$$\delta_k^{(W)}(t) = F_k' \left[\sum_{j=0}^{n_h} W_{k,j} h_j(t) \right] (y_k(t) - \hat{y}_k(t|\theta)) . \tag{2.195}$$

Similarly, the gradient for an input-to-hidden layer weights is given by

$$G(w_{j,l}) = \sum_{t=1}^{N} \varphi_l(t) f_j' \left[\sum_{l=0}^{n_\varphi} w_{j,l} \varphi_l(t) \right] \sum_{k=1}^{n_y} W_{k,j} \delta_k^{(W)}(t)$$

$$= \sum_{t=1}^{N} \varphi_l(t) \delta_j^{(w)}(t) , \tag{2.196}$$

where

$$\delta_j^{(w)}(t) = f_j' \left[\sum_{l=0}^{n_\varphi} w_{j,l} \varphi_l(t) \right] \sum_{k=1}^{n_y} W_{k,j} \delta_k^{(W)}(t) . \tag{2.197}$$

It is straightforward to generalize the algorithm to networks containing more than two layers. The procedure is simply continued backwards layer by layer while back-propagating "deltas" calculated by (2.197).

If one is interested in the partial derivative, $\phi(t)$, only, this is obtained quite easily from (2.194) and (2.196). The prediction error is omitted from (2.195), and the back-propagation is then carried out separately for each output.

Computing the search direction. To determine the search direction in a Gauss-Newton, Levenberg-Marquardt, or Newton algorithm it is necessary to solve a linear system of equations. Since the Gauss-Newton Hessian is always positive definite (and symmetric), the equations can be solved efficiently by using the Cholesky factorization. With Cholesky factorization a lower triangular matrix L is obtained with the properties

$$H = LL^T$$

$$\updownarrow$$

$$
\begin{bmatrix}
h_{11} & h_{21} & \dots & h_{p1} \\
h_{21} & h_{22} & \dots & h_{p2} \\
\vdots & & & \\
h_{p1} & h_{p2} & \dots & h_{pp}
\end{bmatrix}
=
\begin{bmatrix}
l_{11} & & & \\
l_{21} & l_{22} & & 0 \\
\vdots & & \ddots & \\
l_{p1} & l_{p2} & \dots & l_{pp}
\end{bmatrix}
\begin{bmatrix}
l_{11} & l_{21} & \dots & l_{p1} \\
 & l_{22} & & l_{p2} \\
0 & & \ddots & \vdots \\
 & & & l_{pp}
\end{bmatrix}.
\quad (2.198)
$$

The Cholesky factor can be found by application of the algorithm described in Table 2.2. See for example Bierman (1977) and Grewal and Andrews (1993).

Table 2.2. Algorithm for Cholesky factorization.

for $k = 1$ to p
$$l_{kk} = \left[h_{kk} - \sum_{j=1}^{k-1} l_{kj}^2 \right]^{1/2}$$
for $i = k + 1$ to p
$$l_{ik} = \frac{1}{l_{kk}} \left[h_{ik} - \sum_{j=1}^{k-1} l_{ij} l_{kj} \right]^{1/2}$$
end
end

After computation of the Cholesky factor, the search direction can be determined in a two stage procedure employing simple forward and back substitutions. First solve

$$Lf' = -G \qquad (2.199)$$

for f' and then solve

$$L^T f = f' \tag{2.200}$$

for the search direction f. If Cholesky factorization is attempted on a singular matrix or a matrix that is indefinite (which might occur in the Newton method), it will fail. Either due to division by 0 or because the argument to the square root is negative. Thus, the Cholesky factorization is sometimes used for checking whether or not a symmetric matrix is positive definite.

Implementation of the covariance update in a recursive Gauss-Newton algorithm. When applying one of the recursive Gauss-Newton variations, one should be aware that numerical problems eventually occur, when the covariance matrix update is implemented directly as specified in (2.140). The problems will not become evident right away, but typically after $\sim 10^5 - 10^6$ iterations. As a remedy it is in adaptive control common to instead update a factorization of the covariance matrix to guarantee that it remains symmetric and positive definite. See for example Åström and Wittenmark (1995) and Ljung and Söderström (1983). A resetting of the covariance matrix to a diagonal matrix every once in a while is another solution, and it is simpler from an implementation perspective. In relation to off-line neural network training, this is also a quite reasonable solution to the numerical problem.

Multiple inputs. It occurs frequently that the system to be identified has more than one input. It need not be the case that all inputs are control inputs. Some of them might for instance be disturbances that are measured. The model structure selection obviously becomes more complicated in this case as it is necessary to include an additional signal in the regression vector. However, the training methods discussed in this section can be used without further modification.

Multiple outputs. The multi-output case can be somewhat more involved, but it depends on the level of ambition. There are different approaches one can choose between. The simplest is to identify a separate model for each output. In this case, the training methods presented already can be applied directly; it is only necessary to include a number of past values for *each* output in the regression vector. A better strategy is to treat the model as a whole. If the variance of the noise on each output is very different and the cross-correlation is significant, the following criterion should be used for training:

$$V_N(\theta, Z^N) = \frac{1}{2N} \sum_{t=1}^{N} \varepsilon^T(t, \theta) \Gamma^{-1} \varepsilon(t, \theta) \tag{2.201}$$

where Γ is the covariance matrix for the noise, $\Gamma = \mathbf{E}\left\{ e(t) e^T(t) \right\}$. The estimate obtained by minimization of (2.201) corresponds to the maximum likelihood estimate under the assumption that the noise is Gaussian distributed and Γ is known. For use in the Levenberg-Marquardt method, the expressions for gradient and Gauss-Newton Hessian must be changed to

$$G(\theta) = V_N'(\theta, Z^N) = \frac{1}{N} \sum_{t=1}^{N} \psi(t,\theta)\Gamma^{-1}\left[y(t) - \hat{y}(t|\theta)\right] \qquad (2.202)$$

$$R(\theta) = \frac{1}{N} \sum_{t=1}^{N} \psi(t,\theta)\Gamma^{-1}\psi^T(t,\theta) . \qquad (2.203)$$

The derivative of the outputs with respect to the weights is now a matrix instead of a vector

$$\psi(t,\theta) = \frac{\mathrm{d}\hat{y}^T(t|\theta)}{\mathrm{d}\theta} . \qquad (2.204)$$

For the sake of completeness it should also be mentioned that the recursive Gauss-Newton algorithm must be modified to (only the version with exponential forgetting is shown here)

$$K(t) = P(t-1)\psi(t)\left[\Gamma^{-1} + \psi(t)P(t-1)\psi(t)\right]^{-1} \qquad (2.205)$$

$$\theta(t) = \theta(t-1) + K(t)\Gamma^{-1}\left[y(t) - \hat{y}(t|\theta(t-1))\right] \qquad (2.206)$$

$$P(t) = \left[P(t-1) - K(t)\psi(t)P(t-1)\right]\frac{1}{\lambda} . \qquad (2.207)$$

The covariance matrix is of course symmetric and positive definite and thus it should never be a problem to factorize it as $\Gamma^{-1} = \Sigma^T\Sigma$. For simplicity, one may choose to consider the transformed output signal $\bar{y}(t) = \Sigma y(t)$, and the unweighted criterion (Fog et al., 1995)

$$V_N(\theta, Z^N) = \frac{1}{2N} \sum_{t=1}^{N} \left[\bar{y}(t) - \hat{y}(t|\theta)\right]^T \left[\bar{y}(t) - \hat{y}(t|\theta)\right] . \qquad (2.208)$$

If desired, the hidden-to-output layer weights and the (concerned) input-to-hidden layer weights can after the training be rescaled so that the network can work directly on unscaled data.

The covariance matrix is unknown in practice and thus it has to be estimated simultaneously with the weights. The relaxation procedure shown in Table 2.3 is commonly used. It is known as *Iterated Generalized Least Squares (IGLS)*.

It is straightforward to add a regularization term to the criterion. This is done exactly as in the single-output case.

Stopping criteria. In order to terminate the training session automatically it is desirable to develop some suitable stopping criteria. A stopping criterion is typically a scalar variable indicating that nothing significantly is gained by continuing the training. One of the difficulties in selecting a proper stopping

Table 2.3. Iterated generalized least squares (IGLS) algorithm.

1: Initialize the weights, $\theta^{(0)}$, set $j = 1$ and set

$$\Gamma^{(0)} = I .$$

2: Train the network to the minimum using $\Gamma^{(j-1)}$ to obtain $\theta^{(j)}$.
3: Estimate the covariance matrix by

$$\hat{\Gamma}^{(j)} = \frac{1}{N} \sum_{t=1}^{N} \varepsilon(t, \theta^{(j)}) \varepsilon(t, \theta^{(j)})^T .$$

4: Stop if the covariance matrix has converged. Otherwise set $j = j+1$ and go to *Step 2*.

---

criterion is that the nature of the minimization problem sometimes can lead one into believing that the network is properly trained although it is not. For this reason it is usually desirable to apply more than one criterion. Some frequently used stopping criteria are listed below.

Maximum number of iterations: Although this is primarily a criterion indicating the "patience threshold", it is not unreasonable to use it for determining when the weights have converged satisfactory, either. The class of optimization problems is here confined strictly to training of two-layer perceptron networks as models of dynamic systems. This is not a particularly broad class of problems and thus the optimization procedure will behave quite similarly from application to application. One will therefore quickly acquire a good feeling for the number of iterations that a certain training algorithm needs to run at the most. For instance, to reach the minimum with high accuracy it is very unlikely that it is necessary to run more than $5 - 600$ iterations of the proposed Levenberg-Marquardt algorithm for a network with $1 - 500$ weights.

Upper bound for the gradient: In the minimum, $\theta = \hat{\theta}$ the gradient of the criterion should be zero, $G(\hat{\theta}) = 0$. For an iterative minimization method this will never be reached exactly in a finite number of iterations. Alternatively, one can check that some norm of the gradient (e.g., the Euclidian norm) is below a certain value

$$| G(\theta^{(i)}) | \leq \epsilon . \tag{2.209}$$

Unfortunately, it can be difficult to select ϵ since it depends heavily on the magnitude of the weights, θ, and the criterion, $V_N(\theta, Z^N)$, what an adequate value is. Dennis and Schnabel (1983) elaborate further on this subject. It should be emphasized that it is often essential that the gradient is extremely small. Recall for example that it was assumed to be zero in derivation of the final prediction error estimate, discussed in Section 2.4.2. The same assumption will also be made in the derivation of automatic model structure selection techniques treated in Section 2.6.

Upper bound for weight change: Another stopping criterion is based on measuring the weight change between two iterations. If the biggest weight change, $\max_k\{\theta_k^{(i+1)} - \theta_k^{(i)}\}$, is below a certain value one should stop the training.

Upper bound for criterion: It is frequently seen in neural network packages that the user is allowed to specify a value of the criterion below which the training is stopped (Demuth and Beale, 1998). However, in system identification applications it is rare that one knows in advance what an appropriate value should be because the noise variance is typically unknown.

Lower bound for trust region: For neural network models it often occurs that the criterion approximation made by the training algorithm is poor when close to a minimum (Gorodkin et al., 1993). When the Levenberg-Marquardt algorithm is used, one may therefore risk that the trust region is dramatically reduced as the minimum is approached. This may eventually lead to numerical problems since the Levenberg-Marquardt factor, λ, consequently is increased. Thus, it is necessary to introduce a maximum value of λ as an additional stopping criterion.

Early stopping: Section 2.4.2 treated the overfitting phenomenon, and it was shown that it is sometimes better to terminate the training session *before* the minimum of the criterion has been reached. The previously mentioned stopping criteria cannot be used in this case. Instead, it is necessary that an additional data set is available for estimating when the generalization error is smallest.

2.4.4 Section Summary

This section dealt with techniques for neural network training, with special emphasis on using the neural networks for modelling of dynamic systems. The suggested approach was called the *prediction error method* as the network was trained to provide the best possible one-step ahead predictions in a mean square sense.

Criterion. The criterion or *cost function* is a function that in a simple fashion assesses the inaccuracy of a given neural network model. The objective of the training is thus to pick the particular model in the selected model structure that has the smallest value of the criterion. The basic type of criterion presented was based on the mean square error between observed outputs and model predictions. To improve generalization it was recommended to include a regularization term for penalizing model complexity.

Training methods. A training method is an optimization algorithm for finding the minimum of the criterion. It is an iterative search algorithm, which means that a minimum is located by taking a sequence of steps each based on local information about the criterion. The different training methods deviated in the way *search direction* and *step size* are chosen. Several methods were presented:

First-order methods: This class of methods deploy gradient information about the criterion only. The gradient or *steepest decent* method was described. A particular implementation of the method was called the *back-propagation algorithm.*

Second-order methods: This class of methods also make use of the second-order derivative (the Hessian) or some approximation of it. The following methods were covered: the Newton method, the quasi-Newton method, the Gauss-Newton method, the pseudo-Newton method, and the Levenberg-Marquardt method. It was recommended to use the latter as it provides a rapid convergence and generally is a very robust algorithm.

Recursive methods: Sometimes it is relevant to train the neural network on-line, while the data is collected. In certain control designs on-line training is an integral part of the design. In this case the *batch algorithms* mentioned above demands too many computations as they use the entire data set for each optimization step. The recursive methods use only the most recent input-output pair. Methods based on the gradient method and the Gauss-Newton method were discussed.

Local minima. A point of caution about training is that the criterion will have a number of local minima. The training methods will generally only find one of these, which one depends on the initialization of the weights. One will typically deal with this problem by training the network a couple of times, each time starting with a different (random) initialization of the weights.

Generalization. Although the network has been trained to a very small value of the criterion, the model need not be particularly good. A good performance on the training set does *not* automatically imply that the model generalizes well to new inputs. In particular, it was shown that if the model structure was too large (contained too many weights) it lead to overfitting. That is, the noise in the training set was also modelled. The *average generalization error* was introduced as a quantity assessing a given model structure. An estimate of it called "Akaike's Final Prediction Error (FPE)" was also presented. It was discussed that the average generalization error could be decomposed into a *bias error* component and a *variance error* component.

Regularization. It was shown that one way of controlling the average generalization error was to extend the criterion with a term called *regularization by simple weight decay*. The weight decay reduces the variance error at the expense of a higher bias error. The goal is to find a suitable compromise, where the sum of the two errors is small.

Implementation issues. Various aspects of the implementation were discussed; e.g., how to compute the gradients for the different model structures, how to decide when to terminate the training, and how to handle systems with multiple inputs and outputs.

2.5 Validation

In the validation stage, the estimated model is evaluated to clarify if it represents the underlying system adequately. Ideally, the validation should be performed in accordance with the intended use of the model. As it turns out this is often rather difficult to obey. For instance, if the intention is to use the model for designing a control system, the validation ought to imply that a controller was designed and its performance tested in practice. For most applications this level of ambition is somewhat high, and it is common to apply a series of simple "standard" tests instead that each concentrate on investigating particular properties of the model. Although this is less ideal, it is good as a preliminary validation to quickly exclude really poor models.

This section is devoted to survey different such tests that are relevant to neural network models of dynamic systems. Some tests are based on a statistical foundation while others are of a more heuristic nature. Although the present section is entitled *validation*, some means for comparing different models to one another are provided as well. These quantities cannot directly be used for deciding whether or not a particular model should be accepted, but they are useful in relation to the model structure selection issue, which will be further dealt with in Section 2.6.

Most of the tests require a set of data that was not used during training. Such a data set is commonly known as a *test* or *validation set*, and the series of tests based upon it are referred to as *cross-validation*, see e.g., Toussaint (1974) for a general discussion and Hansen and Salamon (1990) for applications in the context of neural networks. It is desirable that the test set satisfies the same demands as the training set regarding representation of the entire operating range. Nevertheless, a small data set is still far better than nothing.

The topics covered in this section have been divided into the following three groups:

Evaluation of the residuals: This includes tests for correlation with different combinations of past residuals and data.

Estimation of the average generalization error: Although they are somewhat unreliable, estimates of the average generalization error can provide a useful picture of the model's ability to predict. In this sense, they are particularly applicable when the available amount of data is so small that one may find it difficult to spare any data for a test set. However, the primary application of the estimates is in relation to model structure selection.

Visualization of the network model's ability to predict: This includes graphical comparison of predictions and observed outputs, multistep ahead predictions, and estimation of prediction intervals.

2.5.1 Looking for Correlations

If the residuals (i.e., the prediction errors) contain no information about past residuals or about the dynamics of the system, it is likely that all information has been extracted from the training set and that the model approximates the system well. To investigate this, one should in principle check if the residuals are uncorrelated with all linear and nonlinear combinations of past data. Such a test is of course completely unrealistic to carry out in practice; thus, it is common to consider only a few wisely chosen auto- and cross-correlation functions. This has been the subject of investigation in several papers of Billings and co-workers, e.g., Billings and Voon (1986), Billings et al. (1992), and Billings and Zhu (1994). The idea is to calculate and visualize sampled correlation functions, since the actual correlation functions are not available. Common to the proposed sampled correlation functions is that they converge in distribution to a Gaussian distribution with zero mean and variance $1/N$ under the hypothesis that the true system has in fact been identified.

Some of the correlation functions recommended in Billings and Voon (1986), Billings et al. (1992), and Billings and Zhu (1994) are listed below:

$$\hat{r}_{\varepsilon\varepsilon}(\tau) = \frac{\sum_{t=1}^{N-\tau} \left(\varepsilon(t,\hat{\theta}) - \bar{\varepsilon} \right) \left(\varepsilon(t-\tau,\hat{\theta}) - \bar{\varepsilon} \right)}{\sum_{t=1}^{N} \left(\varepsilon(t,\hat{\theta}) - \bar{\varepsilon} \right)^2} = \begin{cases} 1, \tau = 0 \\ 0, \tau \neq 0 \end{cases} \quad (2.210)$$

$$\hat{r}_{u\varepsilon}(\tau) = \frac{\sum_{t=1}^{N-\tau} \left(u(t) - \bar{u} \right) \left(\varepsilon(t-\tau,\hat{\theta}) - \bar{\varepsilon} \right)}{\sqrt{\sum_{t=1}^{N} \left(u(t) - \bar{u} \right)^2 \sum_{t=1}^{N} \left(\varepsilon(t,\hat{\theta}) - \bar{\varepsilon} \right)^2}} = 0, \quad \forall \tau \quad (2.211)$$

$$\hat{r}_{u^2\varepsilon^2}(\tau) = \frac{\sum_{t=1}^{N-\tau}[u^2(t) - \overline{u^2}][\varepsilon^2(t - \tau, \hat{\theta}) - \overline{\varepsilon^2}]}{\sqrt{\sum_{t=1}^{N}[u^2(t) - \overline{u^2}]^2 \sum_{t=1}^{N}[\varepsilon^2(t, \hat{\theta}) - \overline{\varepsilon^2}]^2}} = 0, \quad \forall \tau \quad (2.212)$$

$$\hat{r}_{u^2\varepsilon}(\tau) = \frac{\sum_{t=1}^{N-\tau}[u^2(t) - \overline{u^2}][\varepsilon(t - \tau, \hat{\theta}) - \overline{\varepsilon}]}{\sqrt{\sum_{t=1}^{N}[u^2(t) - \overline{u^2}]^2 \sum_{t=1}^{N}[\varepsilon(t, \hat{\theta}) - \overline{\varepsilon}]^2}} = 0, \quad \forall \tau \quad (2.213)$$

$$\hat{r}_{\varepsilon\beta}(\tau) = \frac{\sum_{t=1}^{N-\tau}[\varepsilon(t, \hat{\theta}) - \overline{\varepsilon}][\beta(t - \tau - 1) - \overline{\beta}]}{\sqrt{\sum_{t=1}^{N}[\varepsilon(t, \hat{\theta}) - \overline{\varepsilon}]^2 \sum_{t=1}^{N}[\beta(t) - \overline{\beta}]^2}} = 0,, \quad \tau \geq 0 \quad (2.214)$$

$$\hat{r}_{\alpha\varepsilon^2}(\tau) = \frac{\sum_{t=1}^{N-\tau}[\alpha(t - \tau) - \overline{\alpha}][\varepsilon^2(t - \tau, \hat{\theta}) - \overline{\varepsilon^2}]}{\sqrt{\sum_{t=1}^{N}[\alpha(t) - \overline{\alpha}]^2 \sum_{t=1}^{N}[\varepsilon^2(t, \hat{\theta}) - \overline{\varepsilon^2}]^2}} = \begin{cases} k, \ \tau = 0 \\ 0, \ \tau \neq 0 \end{cases} (2.215)$$

$$\hat{r}_{\alpha u^2}(\tau) = \frac{\sum_{t=1}^{N-\tau}[\alpha(t - \tau) - \overline{\alpha}][u^2(t - \tau) - \overline{u^2}]}{\sqrt{\sum_{t=1}^{N}[\alpha(t) - \overline{\alpha}]^2 \sum_{t=1}^{N}[u^2(t) - \overline{u^2}]^2}} = 0, \quad \forall \tau, \quad (2.216)$$

where

$$\alpha(t) = y(t)\varepsilon(t, \hat{\theta}) \quad (2.217)$$
$$\beta(t) = u(t)\varepsilon(t, \hat{\theta}) \quad (2.218)$$

and

$$k = \frac{\sqrt{\sum_{t=1}^{N}[\varepsilon^2(t, \hat{\theta}) - \overline{\varepsilon^2}]}}{\sqrt{\sum_{t=1}^{N}[\alpha(t) - \overline{\alpha}]^2}}. \quad (2.219)$$

A bar over the symbols has been introduced to specify the average of a signal

$$\bar{x} = \frac{1}{N}\sum_{t=1}^{N} x(t). \quad (2.220)$$

It is common to check if the functions for lags in the interval $\tau \in [-20, 20]$ are zero within an (asymptotical) 95% confidence interval, i.e., if $-1.96/\sqrt{N} < \hat{r} < 1.96/\sqrt{N}$.

The first two tests, (2.210) and (2.211), are very common means of validation in conventional system identification. If one has decided to restrict the attention to linear model structures, there is not much sense in checking for more subtle correlations. Although there may be nonlinear components hidden in the residuals, one will not be able to eliminate them with a linear model structure.

It should be mentioned that all tests apply equally well to MIMO systems. The correlation functions must then be calculated for each combination of input and output.

2.5.2 Estimation of the Average Generalization Error

In this subsection, two techniques for estimating the average generalization error will be derived. Reliable estimates of the average generalization error are useful for validation purposes, but their primary application is for model structure selection. The estimates are good for rapidly comparing different model structures to decide which one is likely to be the best.

The first estimate to be described is Akaike's Final Prediction Error (FPE) estimate while the second estimate is a variation of Leave-One-Out cross-validation. The FPE estimate was introduced in Section 2.4.2, but not on a form that permitted practical evaluation. An estimate of the noise variance will be derived here to make evaluation possible. The leave-one-out cross-validation scheme is often used in linear regression, and in its basic form it requires a quite serious amount of computations when used with neural network models. However, by application of an approximation called *linear unlearning*, the scheme becomes suitable for neural network models as well. The estimate has the acronym LULOO for *Linear-Unlearning-Leave-One-Out*.

Akaike's Final Prediction Error Estimate. Recalled from Section 2.4.2 that the average generalization error can be estimated by

$$\hat{V}_M = \frac{1}{2}\sigma_e^2 \left(1 + \frac{p}{N}\right) \qquad (2.221)$$

if the network has been trained to minimize an unregularized mean square error criterion. In addition to this estimate Ljung (1999) also provides an estimate of the noise variance

$$\hat{\sigma}_e^2 = 2\frac{N}{N-p}V_N(\hat{\theta}, Z^N) . \qquad (2.222)$$

When this is inserted (2.221) the following popular formula appears

$$\hat{V}_M = \frac{N+p}{N-p}V_N(\hat{\theta}, Z^N) . \qquad (2.223)$$

For the *regularized criterion*, with $D = \alpha I$, the FPE estimate is derived as a function of the noise variance in Sjöberg and Ljung (1995). The result was mentioned in Section 2.4.2. The estimate is easily extended to the case of multiple weight decays (Larsen and Hansen, 1994):

$$\hat{V}_M = \frac{1}{2}\left[\sigma_e^2\left(1 + \frac{p_1}{N}\right) + \gamma\right] , \qquad (2.224)$$

where

$$p_1 = \mathbf{tr}\left\{R[R+D]^{-1}\, R[R+D]^{-1}\right\} \tag{2.225}$$

$$\gamma = \frac{1}{N^2}\, \theta_0^T D \left[R + \frac{1}{N}D\right]^{-1} R \left[R + \frac{1}{N}D\right]^{-1} D\theta_0 \ . \tag{2.226}$$

A new estimate of the noise variance is required in this case since (2.221) is for networks trained without regularization. This is derived in the following. The derivations take the same heuristic approach as Sjöberg and Ljung (1995) used for deriving (2.224).

Expanding the mean square error portion of the criterion (the training error) around the true weights, θ_0, and taking the expectation with respect to $\varphi(t)$ and $e(t)$ gives

$$\mathbf{E}\left\{V_N(\hat{\theta}, Z^N)\right\} \simeq \mathbf{E}\left\{V_N(\theta_0, Z^N)\right\} + \mathbf{E}\left\{(\hat{\theta}-\theta_0)^T V_N'(\theta_0, Z^N)\right\}$$
$$+ \ \frac{1}{2}\mathbf{E}\left\{(\hat{\theta}-\theta_0)^T V_N''(\theta_0, Z^N)(\hat{\theta}-\theta_0)\right\} \ . \tag{2.227}$$

The first term on the right-hand side of this expression obviously equals one half of the noise variance

$$\mathbf{E}\left\{V_N(\theta_0, Z^N)\right\} = \bar{V}(\theta_0) = \frac{1}{2}\sigma_e^2 \ . \tag{2.228}$$

The second term is more difficult. It is useful to approximate the gradient of the criterion by a first-order expansion around the true weights,

$$0 = W_N'(\hat{\theta}, Z^N) \simeq W_N'(\theta_0, Z^N) + (\hat{\theta}-\theta_0)^T W_N''(\theta_0, Z^N) \ , \tag{2.229}$$

which leads to

$$(\hat{\theta}-\theta_0) \simeq -\left[W_N''(\theta_0, Z^N)\right]^{-1} W_N'(\hat{\theta}, Z^N) \ . \tag{2.230}$$

In Sjöberg and Ljung (1995) it is pointed out that the law of large numbers justifies that for large N

$$V_N''(\theta_0, Z^N) \simeq \bar{R} = \mathbf{E}\left\{\psi(t, \theta_0)\psi^T(t, \theta_0)\right\} \tag{2.231}$$

implying that

$$W_N''(\theta_0, Z^N) = V_N''(\theta_0, Z^N) + \frac{1}{N}D \simeq \bar{R} + \frac{1}{N}D \ . \tag{2.232}$$

These results are now inserted in the second term on the right-hand side of (2.227)

$$\mathbf{E}\left\{(\hat{\theta}-\theta_0)^T V_N'(\theta_0, Z^N)\right\}$$

$$\simeq \mathbf{E}\left\{\left(-\frac{1}{N}\sum_{t=1}^N \psi(t,\theta_0)e(t) + \frac{1}{N}D\theta_0\right)^T \left[\bar{R}+\frac{1}{N}D\right]^{-1} V_N'(\theta_0, Z^N)\right\}$$

$$\simeq \mathbf{E}\left\{\left(-\frac{1}{N}\sum_{t=1}^N \psi(t,\theta_0)e(t) + \frac{1}{N}D\theta_0\right)^T \left[\bar{R}+\frac{1}{N}D\right]^{-1}\right.$$

$$\left.\times \left(\frac{1}{N}\sum_{t=1}^N \psi(t,\theta_0)e(t)\right)\right\}$$

$$+ \ \mathbf{E}\left\{\left(\frac{1}{N}\theta_0^T D\right)\left[\bar{R}+\frac{1}{N}D\right]^{-1}\left(\frac{1}{N}\sum_{t=1}^N \psi(t,\theta_0)e(t)\right)\right\}. \qquad (2.233)$$

The noise is independent of the input; thus, it is uncorrelated with all functions of the input. The second term in (2.233) is therefore zero. If the trace operator is used, (2.233) can be reduced to:

$$\mathbf{E}\left\{(\hat{\theta}-\theta_0)^T V_N'(\theta_0, Z^N)\right\}$$

$$\simeq -\mathbf{tr}\left\{\mathbf{E}\left\{\left(\frac{1}{N}\sum_{t=1}^N \psi(t,\theta_0)e(t)\right)\left(\frac{1}{N}\sum_{t=1}^N \psi(t,\theta_0)e(t)\right)^T\right\}\right.$$

$$\left.\times \left[\bar{R}+\frac{1}{N}D\right]^{-1}\right\}$$

$$\simeq -\mathbf{tr}\left\{\frac{\sigma_e^2}{N}\bar{R}\left[\bar{R}+\frac{1}{N}D\right]^{-1}\right\}. \qquad (2.234)$$

Finally, the third term on the right-hand side of (2.227) is evaluated

$$\mathbf{tr}\left\{\mathbf{E}\left\{\left(\hat{\theta}-\theta_0\right)^T \bar{R}\left(\hat{\theta}-\theta_0\right)\right\}\right\} = \mathbf{tr}\left\{\bar{R}\mathbf{E}\left\{\left(\hat{\theta}-\theta_0\right)\left(\hat{\theta}-\theta_0\right)^T\right\}\right\}$$

$$= \mathbf{tr}\left\{\bar{R}P\right\}. \qquad (2.235)$$

The covariance matrix P, which is the expectation of the product of the weight deviations, is determined in Sjöberg and Ljung (1995) for a single weight decay. Using the results in (2.230) and (2.232) the derivations are straightforward to extend to multiple weight decays:

$$P \overset{\triangle}{=} \mathbf{E}\left\{\left(\hat{\theta} - \theta_0\right)\left(\hat{\theta} - \theta_0\right)^T\right\}$$

$$\simeq \frac{\sigma_e^2}{N}\left(\bar{R} + \frac{1}{N}D\right)^{-1}\bar{R}\left(\bar{R} + \frac{1}{N}D\right)^{-1}$$

$$+ \frac{1}{N^2}\left(\bar{R} + \frac{1}{N}D\right)^{-1}D\theta_0\theta_0^T D\left(\bar{R} + \frac{1}{N}D\right)^{-1}. \quad (2.236)$$

This leads to the expression

$$\mathbf{tr}\left\{\bar{R}P\right\} = \frac{\sigma_e^2}{N}\bar{R}\left(\bar{R} + \frac{1}{N}D\right)^{-1}\bar{R}\left(\bar{R} + \frac{1}{N}D\right)^{-1}$$

$$+ \frac{1}{N^2}\theta_0^T D\left(\bar{R} + \frac{1}{N}D\right)^{-1}\bar{R}\left(\bar{R} + \frac{1}{N}D\right)^{-1}D\theta_0^T. \,(2.237)$$

$\bar{R}$ is of course unknown and is replaced by the Gauss-Newton Hessian evaluated in the (biased) minimum $R \overset{\triangle}{=} R(\hat{\theta})$. Recognizing p_1 and γ in (2.237) and by inserting (2.228), (2.234), and (2.237) in (2.227), the following result is obtained

$$2\mathbf{E}\left\{V_N(\hat{\theta}, Z^N)\right\} \simeq \sigma_e^2 - 2\frac{\sigma_e^2}{N}\mathbf{tr}\left\{R\left(R + \frac{1}{N}D\right)^{-1}\right\}$$

$$+ \frac{\sigma_e^2}{N}p_1 + \gamma. \quad (2.238)$$

Defining p_2 as

$$p_2 = \mathbf{tr}\left\{R\left(R + \frac{1}{N}D\right)^{-1}\right\} \quad (2.239)$$

(2.238) can also be written in the form

$$2\mathbf{E}\left\{V_N(\hat{\theta}, Z^N)\right\} \simeq \sigma_e^2\left(1 + \frac{p_1 - 2p_2}{N}\right) + \gamma \quad (2.240)$$

which leads to the following estimate of the noise

$$\hat{\sigma}_e^2 = \frac{2NV_N(\hat{\theta}, Z^N) - N\gamma}{N + p_1 - 2p_2}. \quad (2.241)$$

According to the previous discussion concerning the interpretation of p_1, for $D = \alpha I$ it is not unreasonable to use

$$p_2 = \mathbf{tr}\left\{R\left(R + \frac{1}{N}D\right)^{-1}\right\} = \sum_{i=1}^{p}\frac{\delta_i}{\delta_i + \frac{\alpha}{N}} \simeq p_1. \quad (2.242)$$

If γ is discarded and the variance estimate is inserted in (2.224), the following FPE estimate(s) is achieved

$$\hat{V}_M = \frac{N + p_1}{N + p_1 - 2p_2} V_N(\hat{\theta}, Z^N) \simeq \frac{N + p_1}{N - p_1} V_N(\hat{\theta}, Z^N) . \qquad (2.243)$$

γ is a positive quantity. Ignoring it thus has the effect that (2.224) becomes too small while (2.241) becomes too big. By the way in which these two expressions are combined into (2.243), the effect is obviously amplified rendering the FPE estimate too small. If one insists on including γ in the calculation for improving the accuracy of the estimate, the true weights in θ_0 must then be substituted for the biased estimates in $\hat{\theta}$. A more accurate FPE, which contains some additional terms, is reported in Larsen and Hansen (1994).

The estimate of the average generalization error has been derived under the assumption that the true system is contained in the model structure, $\mathcal{S} \in \mathcal{M}$, but in practice it often seems to work well regardless of the fact that this condition is not satisfied.

The LULOO Estimate. An alternative technique for estimating the average generalization error is by leave-one-out cross-validation. The principle behind leave-one-out is to simulate that a test set is available. The principle is quite intuitive: a network is trained on the entire data set except for one input-output pair $Z^N \setminus \{\varphi(t), y(t)\}$. The prediction error is evaluated on the pair $\{\varphi(t), y(t)\}$ and the procedure is then repeated for another t. When the prediction errors have been calculated for all $1 \le t \le N$, the average generalization error is estimated as (one half) their mean square error. If $\hat{\theta}_t$ denotes the minimum of the network trained on $Z^N \setminus \{\varphi(t), y(t)\}$ the LOO estimate can be formulated as (Efron and Tibshirani, 1993):

$$\hat{V}_M = \frac{1}{2N} \sum_{t=1}^{N} \left[y(t) - \hat{y}(t|\hat{\theta}_t) \right]^2 = \frac{1}{2N} \sum_{t=1}^{N} \varepsilon^2(t, \hat{\theta}_t) . \qquad (2.244)$$

The estimate can only be used with non-recursive networks as the regression vector, $\varphi(t)$, would otherwise depend on past outputs of the model. In this case one cannot just remove a single input-output pair from the data set. Thus, the basic leave-one-out scheme is restricted to NNFIR and NNARX model structures.

Since leave-one-out cross-validation implies that N networks are trained to their minimum, the scheme must be considered unrealistic except for small data sets. Usually a short-cut is made in the implementation, however, to avoid an exhaustive search for a global minimum: The network is trained on the entire data set a couple of times, and the lowest minimum found is stored. This minimum is then used as a starting point for the training carried on the N reduced data sets. Typically only a few iterations are necessary as the relocation of the minimum generally is minor when only one input-output pair is removed from the data set. There is, however, no guarantee that the global minimum for the reduced data set is close to the global minimum.

Even with the mentioned short-cut, the required amount of computations is rather high. In Hansen and Larsen (1996) the concept of *linear unlearning* is introduced to circumvent the many retrainings. In virtue of linear unlearning, which is based on a quadratic approximation to the criterion of fit, a reasonable estimate of the average generalization error can be achieved by simple means. Experience has shown that the accuracy of this *Linear Unlearning Leave-One Out (LULOO) estimate* typically is comparable to the FPE estimate. The LULOO estimate is, however, generally not as accurate as true leave-one out. The LULOO estimate is derived in the following for networks trained according to the regularized mean square error criterion.

Let Z_t^{N-1} denote the reduced data set obtained by leaving out input-output pair number t

$$Z_t^{N-1} = Z^N \setminus \{\varphi(t), y(t)\} , \qquad (2.245)$$

and let $\hat{\theta}_t$ specify the minimum of the regularized criterion:

$$W_{N-1}(\theta, Z_t^{N-1}) = W_N(\theta, Z^N) - \frac{1}{2N}[y(t) - \hat{y}(t|\theta)]^2 . \qquad (2.246)$$

To conform to the notation used previously, the right-hand side should in principle be normalized by the factor $N/(N-1)$. This is not so important since the minimizer is independent of such a normalization.

Introduce the short-hand notation

$$H_t = W_{N-1}''(\hat{\theta}, Z_t^{N-1}) \qquad (2.247)$$

$$g_t = \left.\frac{\partial[y(t) - \hat{y}(t|\theta)]^2}{\partial\theta}\right|_{\theta=\hat{\theta}} = -2\psi(t, \hat{\theta})\varepsilon(t, \hat{\theta}) \qquad (2.248)$$

A linear approximation of $W_{N-1}'(\theta, Z_t^{N-1})$ around $\hat{\theta}$ gives

$$0 = W_{N-1}'(\hat{\theta}_t, Z_t^{N-1}) \simeq W_{N-1}'(\hat{\theta}, Z_t^{N-1}) + (\hat{\theta} - \hat{\theta}_t)^T H_t \qquad (2.249)$$

$$= W_N'(\hat{\theta}, Z^N) - \frac{1}{2N}g_t + (\hat{\theta} - \hat{\theta}_t)^T H_t$$

$$= -\frac{1}{2N}g_t + (\hat{\theta} - \hat{\theta}_t)^T H_t , \qquad (2.250)$$

or

$$(\hat{\theta} - \hat{\theta}_t) = \frac{1}{2N}g_t H_t^{-1} . \qquad (2.251)$$

This result can be used to obtain an approximate expression for the squared errors entering (2.244):

$$\varepsilon^2(t, \hat{\theta}_t) \simeq \varepsilon^2(t, \hat{\theta}) + (\hat{\theta} - \hat{\theta}_t)^T g_t$$

$$\simeq \varepsilon^2(t, \hat{\theta}) + \frac{1}{2N} g_t^T H_t^{-1} g_t$$

$$= \varepsilon^2(t, \hat{\theta}) \left(1 + \frac{2}{N} \psi^T(t, \hat{\theta}) H_t^{-1} \psi(t, \hat{\theta})\right) \qquad (2.252)$$

which, when inserted in (2.244), leads to an estimate of the average generalization error that does not require retraining

$$\hat{V}_M = \frac{1}{2N} \sum_{t=1}^{N} \varepsilon^2(t, \hat{\theta}) \left(1 + \frac{2}{N} \psi^T(t, \hat{\theta}) H_t^{-1} \psi(t, \hat{\theta})\right) . \qquad (2.253)$$

As discussed in Section 2.4.1, the full Hessian is difficult and time-consuming to calculate. As it is necessary to evaluate the Hessian N times, calculating (2.253) is therefore a quite demanding task. A remedy for this is to invoke the Gauss-Newton approximation of the Hessian as this is generally a good approximation in the minimum.

$$W''(\hat{\theta}, Z^N) \simeq H = \frac{1}{N} \sum_{t=1}^{N} \psi(t, \hat{\theta}) \psi^T(t, \hat{\theta}) + \frac{1}{N} D . \qquad (2.254)$$

By application of the matrix inversion lemma

$$(A^{-1} + BCD)^{-1} = A - AB(C^{-1} + DAB)^{-1} DA \qquad (2.255)$$

to

$$H_t = H - \frac{1}{N} \psi(t, \hat{\theta}) \psi^T(t, \hat{\theta}) \qquad (2.256)$$

the following expression is obtained:

$$H_t^{-1} = H^{-1} + \frac{H^{-1} \psi(t, \hat{\theta}) \psi^T(t, \hat{\theta}) H^{-1}}{N - \psi^T(t, \hat{\theta}) H^{-1} \psi(t, \hat{\theta})} . \qquad (2.257)$$

When this is inserted in (2.253), an estimate that is both simple and inexpensive to calculate appears

$$\hat{V}_M = \frac{1}{2N} \sum_{t=1}^{N} \left[\varepsilon^2(t, \hat{\theta}) \frac{N + \psi^T(t, \hat{\theta}) H^{-1} \psi(t, \hat{\theta})}{N - \psi^T(t, \hat{\theta}) H^{-1} \psi(t, \hat{\theta})} \right] . \qquad (2.258)$$

As pointed out in Hansen and Larsen (1996), the expression has some similarities with the FPE estimate. Interpreting the quantity $\left[\psi^T(t, \hat{\theta}) H^{-1} \psi(t, \hat{\theta})\right]$ as the effective number of weights for input-output pair t, the expression can be regarded as being an "example-based" FPE estimate. A related test error estimator has been proposed in Wahba (1990) and a related unlearning approach based on retraining was presented in Moody (1994).

2.5.3 Visualization of the Predictions

The test error and the estimates of average generalization error measure the accuracy of the predictions in terms of a scalar quantity. This quantity gives an overall estimate of the accuracy of the prediction error and will not give information about variation in accuracy among different regimes of the operating range of the system. A simple plot that compares predictions to actual measurements in training and test set can provide a better understanding of these variations. Unless the signal-to-noise ratio is very poor it can show the extent of overfitting as well as possible systematic errors. Although it may appear somewhat hand-waved compared to the correlation tests, the practical value of a visual understanding should not be underestimated. Figure 2.20 and Figure 2.21 demonstrates how one visually can detect if a model is either underparametrized or overparametrized. Figure 2.20 displays the predictions on training and test set for an underparametrized model while Figure 2.21 displays the predictions for an overparametrized model.

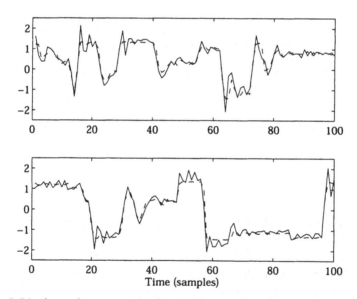

Figure 2.20. An underparametrized network model evaluated on a section of the training set (upper figure) and on a test set (lower figure). The solid line is the observed output and the dashed line is the prediction. The predictions are obviously poor in both cases.

Prediction intervals. In many practical applications one is interested in knowing something about the *reliability* of the predictions for a specific input to the network. In Sørensen et al. (1996) a method is discussed for estimating

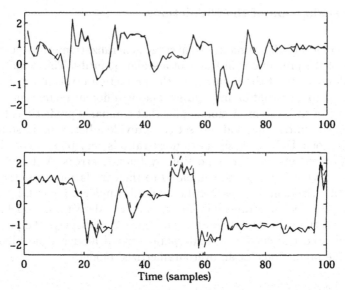

Figure 2.21. An overparametrized network model evaluated on a section of the training set (upper figure) and on a the test set (lower figure). The solid line is the observed output and the dashed line is the prediction. Notice that the training data set is modelled accurately despite the fact that it is affected by noise. However, when the network is evaluated on test data, the predictions are very poor.

error bars for the predictions; so-called prediction intervals. The method is based on the derivations in Section 2.5.2 and requires that the condition $S \in \mathcal{M}$ is satisfied.

Consider the variance of the prediction error conditioned on the regression vector, $\varphi(t)$,

$$\sigma_p^2(t) = \mathbf{E}\left\{ \varepsilon^2(t,\hat{\theta}) \Big| \varphi(t) \right\} = \mathbf{E}\left\{ \left[y(t) - \hat{y}(t|\hat{\theta}) \right]^2 \Big| \varphi(t) \right\} . \qquad (2.259)$$

If the variance was known and the prediction errors were assumed to be Gaussian distributed it would then be possible to calculate a confidence interval for the predictions: $y(t) \in \left[\hat{y}(t|\hat{\theta}) - c\sigma_p; \ \hat{y}(t|\hat{\theta}) + c\sigma_p \right]$, where c is selected in accordance with the desired confidence interval.

Obviously, the variance, σ_p^2, is composed of two parts: a contribution due to the noise and a contribution due to "non-optimal" predictions

$$\sigma_p^2(t) = \mathbf{E}\left\{ \left[y(t) - \hat{y}(t|\hat{\theta}) \right]^2 \Big| \varphi(t) \right\}$$

$$= \sigma_e^2 + \mathbf{E}\left\{ \left[g_0[t,\theta_0] - \hat{y}(t|\hat{\theta}) \right]^2 \Big| \varphi(t) \right\} . \qquad (2.260)$$

Under the assumption that $\mathcal{S} \in \mathcal{M}$, g_0 can be approximated by the following expansion

$$g_0[t, \theta_0] \simeq \hat{y}(t, \hat{\theta}) + (\theta_0 - \hat{\theta})^T \psi(t, \hat{\theta}) \qquad (2.261)$$

which, when inserted in (2.260), gives

$$\sigma_p^2(t) = \sigma_e^2 + \psi^T(t, \hat{\theta}) \mathbf{E} \left\{ (\theta_0 - \hat{\theta})(\theta_0 - \hat{\theta})^T \right\} \psi(t, \hat{\theta}) . \qquad (2.262)$$

The covariance matrix can be estimated according to (2.236), but it is also possible to estimate it as an intermediate result of the LULOO procedure (it is ignored that $\hat{\theta}$ and $\hat{\theta}_t$ are biased estimates when regularization is used)

$$\mathbf{E} \left\{ (\theta_0 - \hat{\theta})(\theta_0 - \hat{\theta})^T \right\} \simeq \sum_{t=1}^{N} (\hat{\theta}_t - \hat{\theta} - \bar{\theta})(\hat{\theta}_t - \hat{\theta} - \bar{\theta})^T , \qquad (2.263)$$

where $\bar{\theta}$ specifies the sample mean

$$\bar{\theta} = \frac{1}{N} \sum_{t=1}^{N} (\hat{\theta}_t - \hat{\theta}) . \qquad (2.264)$$

The derivation of (2.263) is somewhat involved and it is referred to Seber and Wild (1989) for further details.

The noise variance, σ_e^2, can be estimated according to the formula derived as a part of the final prediction error estimate in the previous subsection (2.241). However, an estimate can also be derived as a spin-off from the LULOO procedure

$$\mathbf{E} \left\{ V_N(\theta_0, Z^N) \right\} \simeq \mathbf{E} \left\{ V_N(\hat{\theta}, Z^N) \right\}$$
$$+ \frac{1}{2} \mathbf{E} \left\{ \left(\hat{\theta} - \theta_0 \right)^T H \left(\hat{\theta} - \theta_0 \right) \right\} . \qquad (2.265)$$

The left-hand side is recognized as the noise variance. Let the first term on the right-hand side be substituted for the only observation of it. When applying the trace operator to the second term on the right-hand side, the following estimate appears

$$\hat{\sigma}_e^2(t) \simeq 2V_N(\hat{\theta}_0, Z^N) + \mathbf{tr} \left\{ H \mathbf{E} \left\{ \left(\hat{\theta} - \theta_0 \right) \left(\hat{\theta} - \theta_0 \right)^T \right\} \right\}$$
$$\simeq 2V_N(\hat{\theta}_0, Z^N) + \mathbf{tr} \left\{ H \sum_{t=1}^{N} (\hat{\theta}_t - \hat{\theta} - \bar{\theta})(\hat{\theta}_t - \hat{\theta} - \bar{\theta})^T \right\} . \qquad (2.266)$$

To give an impression of how the error bars can enhance interpretation of the predictions, an example is displayed in Figure 2.22. An NNARX(2,2,1) model

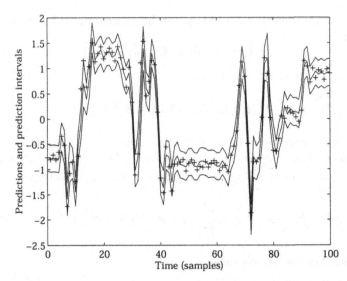

Figure 2.22. An NNARX model evaluated on a test set. A 95% confidence interval has been drawn around the predictions. 'x' is the observed output.

was trained on a data set of 400 samples. It is assumed that the prediction errors are Gaussian distributed.

Many approximations were made to yield the formula for the estimation of prediction intervals; thus, one should not rely too much on the results. The primary use in practice is probably as a tool for pinpointing regimes of the operating range where the predictions are particularly unreliable.

k-step ahead predictions. If the sampling frequency is high compared to the dynamics of the system, a visual inspection of the predictions will usually not reveal possible problems. Two succeeding outputs will always be very close, $y(t) \simeq y(t-1)$, and unless the one-step ahead prediction, $\hat{y}(t|\hat{\theta})$, is a much worse prediction than $y(t-1)$, the prediction error will appear to be very small. In this case, a small prediction error clearly does not imply that the model is good. In case of fast sampling it may thus be a good idea to check that $\hat{y}(t|\hat{\theta})$ is in fact a much better prediction of $y(t)$ than the so-called "naive prediction", $\hat{y}(t|\hat{\theta}) = y(t-1)$. Another useful technique is to inspect the k-step ahead prediction.

The k-step ahead prediction is calculated as the *one*-step ahead prediction, except that predictions substitute observed outputs where none exist, and the future residuals are set to zero. For an NNARX model this k-step ahead prediction takes the form

$$\hat{y}(t+k) \stackrel{\triangle}{=} \hat{y}(t+k|t,\hat{\theta}) = \hat{g}\left[\bar{\varphi}(t+k),\hat{\theta}\right] , \qquad (2.267)$$

where

$$\bar{\varphi}^T(t+k) = [\hat{y}(t+k), \ ... \ , \hat{y}(t+k-\min(k,n)+1),$$
$$y(t) \ ... \ , y(t-\max(n-k,0)),$$
$$u(t-d+k), \ ... \ , u(t-d-m+k)] \ . \quad (2.268)$$

This looks a little complicated, but it is really quite intuitive. The input vector to an NNARX(2,2,1) model is for $k = 1$ (the well known case) to $k = 3$:

$$\bar{\varphi}^T(t+1) = [y(t), \ y(t-1), \ u(t), \ u(t-1)]$$
$$\bar{\varphi}^T(t+2) = [\hat{y}(t+1), \ y(t), \ u(t+1), \ u(t)]$$
$$\bar{\varphi}^T(t+3) = [\hat{y}(t+2), \ \hat{y}(t+1), \ u(t+2), \ u(t+1)] \ .$$

Although the network was not trained with this application in mind (it was trained to provide *one*-step ahead predictions), the k-step ahead predictions will often reveal whether or not the most essential properties have been captured.

An illustration of the alleged problem is found below. For a deterministic system an experiment was conducted by using a high sampling frequency. The *one*-step ahead predictions are displayed in the upper panel of Figure 2.23, while the *ten*-step ahead predictions are displayed in the lower panel. In the latter case, the discrepancy between predictions and observed outputs is significant.

This type of validation will be particularly useful in relation to the predictive control strategies discussed in Section 3.8.

2.5.4 Section Summary

This section provided some means for assessing whether or not a model is able to reproduce data to an acceptable accuracy. In addition, some quantities for comparing different model structures were discussed.

Correlation functions. If all information about the dynamics of the system has been incorporated into the model, the prediction errors should be independent of both control input and of the output signal. It has been suggested that the hypothesis of independence was checked through calculation of the cross-correlation function between functions of the prediction errors and functions of input and output. In principle, the independence should be checked by looking at the correlation between all possible functions of prediction errors and inputs. However, in practice one must rely on a few sensible examples.

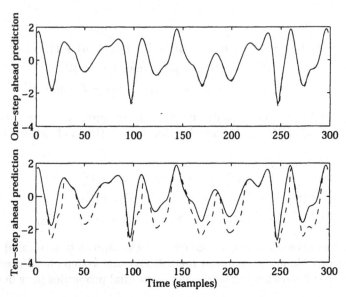

Figure 2.23. Upper panel: The *one*-step ahead predictions. Lower panel: The *ten*-step ahead predictions. The solid lines are the observed outputs and the dashed lines are the predictions.

Estimating average generalization error. The generalization error and the *average* generalization error assess how well model and model structure, respectively, are expected to generalize to arbitrary data. The generalization error can be estimated by evaluating the mean square error on a *test set*; i.e., a data set not used for training. Different estimates of the average generalization errors were provided: *Akaike's Final Prediction Error* estimate (FPE), *Leave-One-Out*, and the simplified *Linear-Unlearning Leave-One-Out*. These estimates are all calculated by using the data in the *training set*.

Visualization of predictions. It was mentioned that a very important part of the validation is to simply inspect plots comparing observed outputs to predictions. In this way one can often detect *underfitting* and *overfitting*. If was illustrated that if the sampling frequency is high compared to the dynamics of the system, the one-step ahead predictions of a poor model can look very accurate when inspected visually. In this case one can look at the k-step ahead prediction instead. The choice $k = \infty$ corresponds to a pure simulation of the system.

A simple method for estimating prediction intervals was derived. These intervals can reveal if, in certain regimes of the operating range, the predictions are particularly poor. This might guide one in performing a new experiment in which these regimes are explored more carefully.

2.6 Going Backwards in the Procedure

The identification procedure depicted in Figure 2.1 has three paths going from
the model validation and back to the previous stages. These paths symbolize
that it is necessary to go back in the procedure if a model is not accepted
immediately. The reason for stepping back in the procedure is two-fold: either
one is dissatisfied with the network model that was just trained, or one may
simply wish to explore if it is possible to find a better model. In this section
it will be discussed what this "stepping back" involves and how it is done in
an intelligent fashion. The attention is primarily concentrated on the path
leading back to the model structure selection. For solving this task in a highly
automated fashion, a class of methods known as pruning algorithms will be
presented.

2.6.1 Training the Network Again

The path leading back to the network training stage in Figure 2.1 covers up
two things:

- Retrain the network with a different initialization of the weights.

- Retrain the network according to another criterion.

The example depicted in Figure 2.19 provides an excellent illustration of the
problem: the generalization ability of the network depends heavily on the
magnitude of the weight decay as well as on the minimum found by the
training method. For a given model structure, one is therefore interested in
determining the weight decay and minimum that result in optimal generalization. As mentioned previously, the trial-and-error test is perhaps the most
reliable strategy for determination of the optimal weight decay and for pointing out the global minimum. However, if a test set is not available it is not
possible to carry out such a test. It might in this case be useful to evaluate
the average generalization error for the trained networks in terms of the FPE
and LULOO estimates. One should be careful not to rely too heavily on these
estimates, though, as they were derived using long series approximations.

As another point of caution it should be emphasized that the global minimum
is not necessarily the best minimum from a generalization error perspective.
If the network is massively overparametrized, it is often advantageous to train
the network to one of the non-global minima (compare to early stopping and
the overtraining phenomenon).

2.6.2 Finding the Optimal Network Architecture

Determination of a model structure encompasses that two fundamental choices are made:

1: The contents of the regression vector.

2: What weights to include in the network architecture (which in turn encompasses selecting a number of hidden units).

To select a sensible model structure, some sort of criterion specifying the goodness of a given structure is required. A natural way to rank different model structures is obviously in terms of their average generalization error. Since the model structure selection is subject to a combinatorial explosion it is impossible to investigate all configurations. In practice one has to be content with a trade-off between doing a fair amount of computations and obtaining an architecture that is reasonably near the optimal.

The common working procedure is to separate the two components of the problem by first selecting a particular structure of the regression vector (NNARX, NNARMAX, NNOE, n past outputs, m past inputs, etc.) and subsequently trying to determine the best possible network architecture for this choice of regressors. If one has access to an unlimited amount of training data the architecture determination becomes less important and one can safely restrict the attention to fully connected networks. For fully connected networks the architecture selection is reduced to a matter of choosing a number of hidden units. The simplest procedure for determination of an adequate number of hidden units is to increase their number gradually while evaluating the test error. When a number of hidden units has been reached above which the gain in generalization is insignificant, the network is accepted without further ado. If, on the other hand, the training set is very limited, it is important that the network architecture is chosen wisely in that it should contain only the most essential weights. The architecture selection is in this case much harder since it will also be difficult to set aside a data set for test purposes. Different attempts have been made to automate the architecture selection for a given set of regressors. Among these attempts are the most popular procedures probably the so-called pruning algorithms. The common strategy in these algorithms is that one initially finds a fully connected network architecture which (in principle) is large enough to describe the system, $S \in M$. Starting with this architecture, the weights are then eliminated one at a time until the optimal architecture has been reached. Some may find it more natural to work in the opposite direction by starting with a small network architecture and then gradually increase it. However, the statistical analysis that one generally has to rely on for ranking different candidate architectures usually requires that the condition $S \in M$ can be assumed to be satisfied. Thus, it is necessary to start with a large network and then shrink it.

Regularization has not yet been mentioned in this subsection despite the fact that its effect is somehow related to choice of network architecture. In both cases the subject of the matter is to "control" the generalization error. If a sensible weight decay parameter is chosen, a subsequent pruning of the network can only be expected to provide a modest improvement in generalization. However, it would be desirable if a pruning of the network could make regularization superfluous whereby the tedious trial-and-error search for a weight decay could be avoided. Unfortunately, pruning algorithms usually cannot perform this kind of miracle. Recall that one of the features of weight decay was its smoothing effect on the criterion. This smoothness is generally an important requirement in order for the pruning algorithms to be successful. Before executing a pruning algorithm it is therefore recommended to first determine a weight decay that is somewhat smaller than the optimal one ($\sim$ a factor $2 - 100$).

What is then a rational procedure for eliminating the weights? Although it has been decided to apply a strategy where only one weight is eliminated at a time, it is still unrealistic to retrain the network to evaluate each of the possible scenarios. What is required is a simple method for ranking the different weights, which in turn can point out the most favorable one to eliminate. The ranking of the weights can be done in different ways. As discussed previously, minimization of the unregularized criterion gives, under mild assumptions, a consistent and asymptotically Gaussian distributed estimate of the weights

$$\hat{\theta} - \theta_0 \in AsN(0, \frac{1}{N}P) , \qquad (2.269)$$

where

$$P \simeq \hat{\sigma}_e^2 \left[\frac{1}{N} \sum_{t=1}^{N} \psi(t, \hat{\theta}) \psi^T(t, \hat{\theta}) \right]^{-1} . \qquad (2.270)$$

Obvious candidates for elimination are naturally those weights $\{\theta_j\}$ for which zero is within a distance of two or three standard deviations from the estimates, $\{\hat{\theta}_j\}$. As a way for ranking these candidates by significance, one can extend this to consider the following ratio that compares a weight to its standard deviation

$$r_j = \frac{|\hat{\theta}_j|}{\sqrt{P_{jj}}} . \qquad (2.271)$$

If the ratio is small, i.e., the variance of the weight is large compared to the magnitude of the estimate, one must expect that an elimination of the weight will cause a large reduction in the variance error portion of the average generalization error. As it turns out, it can be proved that for a network trained according to the *unregularized* criterion, elimination of the weight for which this ratio is smallest actually leads to the maximum decrease in the average generalization error. The idea of eliminating the weights according to (2.271) was suggested by Hassibi and Stork (1993), which introduced the *Optimal Brain Surgeon (OBS)*. Originally the OBS was derived for unregularized mean square error criteria, and it was not directly motivated by the generalization error considerations mentioned above. The extension to the regularized case and the connection to the average generalization error were first treated in Hansen and Pedersen (1994) and Pedersen and Hansen (1995).

Optimal Brain Surgeon (OBS). Recall that for a single output network model trained to its minimum, the FPE estimate of the average generalization error is given by

$$\hat{V}_M = \frac{N + p}{N - p} V_N(\hat{\theta}, Z^N) \,, \tag{2.272}$$

for the unregularized criterion, and

$$\begin{aligned}
\hat{V}_M &= \frac{N + p_1}{N + p_1 - 2p_2} V_N(\hat{\theta}, Z^N) \\
&\simeq \frac{N + p_1}{N - p_1} V_N(\hat{\theta}, Z^N) \\
&\simeq \left(1 + \frac{2p_1}{N}\right) V_N(\hat{\theta}, Z^N)
\end{aligned} \tag{2.273}$$

if a regularization term is included. The last of the three estimates is used in Pedersen and Hansen (1995), which is why it has been included here. In practice one can expect little difference in accuracy between the three estimates. The principle of OBS is to eliminate the weight that gives a maximum decrease in the FPE estimate. Thus, one must first investigate how each of the weights affects the FPE estimate.

Let M_j specify the model structure with the jth weight removed and introduce the following notation

$$\hat{\theta}^{(j)} \triangleq \arg\min_\theta W_N(\theta, Z^N) \text{ subject to } \theta_j = 0 \tag{2.274}$$

$$\delta V^{(j)} \triangleq V_N(\hat{\theta}^{(j)}, Z^N) - V_N(\hat{\theta}, Z^N) \tag{2.275}$$

$$\delta V_M^{(j)} \triangleq V_{M_j} - V_M \,. \tag{2.276}$$

(2.275) specifies the change in training error and is denoted the *saliency* for weight j. (2.276) specifies the change in average generalization error and is denoted the *generalization-based saliency for weight j*. The objective of OBS is in other words to eliminate the weight with smallest generalization-based saliency. From (2.272) it appears that in the unregularized case this is the same as eliminating the weight with the smallest saliency. In the regularized case it is, however, necessary to find the minimizer of

$$\delta \hat{V}_M^{(j)} = \delta V^{(j)} - \frac{2(p_1 - p_1^{(j)})}{N} V_N(\hat{\theta}, Z^N) + \frac{2p_1^{(j)}}{N} \delta V^{(j)}$$

$$\simeq \delta V^{(j)} - \frac{2(p_1 - p_1^{(j)})}{N} V_N(\hat{\theta}, Z^N) , \qquad (2.277)$$

where $p_1^{(j)}$ is the *effective number of weights* in M_j.

Computing the saliencies:

To avoid having to retrain the network in order to determine the saliencies, $\delta V^{(j)}$, and the effective number of weights, $p_1^{(j)}$, a series expansion of the criterion of fit is made around the current minimum

$$\tilde{W}_N(\theta, Z^N) = W_N(\hat{\theta}, Z^N) + \frac{1}{2} \left(\theta - \hat{\theta}\right)^T W_N''(\hat{\theta}, Z^N) \left(\theta - \hat{\theta}\right) . \qquad (2.278)$$

As usual, the Gauss-Newton approximation of the true Hessian is invoked since it is reasonably accurate considering the expansion is made around the minimum

$$W_N''(\hat{\theta}, Z^N) \simeq H = R + \frac{1}{N} D . \qquad (2.279)$$

Elimination of the jth weight corresponds to minimizing (2.278) subject to the constraint

$$\theta_j = 0 \quad \Leftrightarrow \quad \theta^T e_j = 0 \quad \Leftrightarrow \quad e_j^T(\theta - \hat{\theta}) + \hat{\theta}_j = 0 , \qquad (2.280)$$

where e_j is the jth unit vector. The constrained minimum, $\hat{\theta}^{(j)}$, is thus found via the Lagrangian

$$L(\theta, \lambda) = \frac{1}{2} \left(\theta - \hat{\theta}\right)^T H \left(\theta - \hat{\theta}\right) + \lambda \left(e_j^T(\theta - \hat{\theta}) + \hat{\theta}_j\right) . \qquad (2.281)$$

λ is the Lagrange multiplier and is multiplied to the constraint. $W(\hat{\theta}, Z^N)$ in (2.278) is a constant and is thus left out. To find the minimizing pair $\{\hat{\theta}^{(j)}, \lambda^{(j)}\}$ of $L(\theta, \lambda)$ its derivative with respect to λ and θ is found and set to zero:

$$\frac{\partial L(\theta, \lambda)}{\partial \lambda} = e_j^T(\theta - \hat{\theta}) + \hat{\theta}_j = 0 \quad \Leftrightarrow \quad e_j^T(\theta - \hat{\theta}) = -\hat{\theta}_j \qquad (2.282)$$

$$\frac{\partial L(\theta, \lambda)}{\partial(\theta - \hat{\theta})} = H \left(\theta - \hat{\theta}\right) + \lambda e_j = 0$$

$$\Leftrightarrow H^{-1}\lambda e_j = -\left(\theta - \hat{\theta}\right)$$

$$\Leftrightarrow e_j^T H^{-1}\lambda e_j = -e_j^T \left(\theta - \hat{\theta}\right) . \qquad (2.283)$$

Inserting (2.282) in (2.283) gives the Lagrange multiplier

$$\lambda^{(j)} = \frac{\hat{\theta}_j}{[H^{-1}]_{jj}} \, , \tag{2.284}$$

and the constrained minimum now follows immediately from (2.283)

$$\hat{\theta}^{(j)} - \hat{\theta} = -\frac{\hat{\theta}_j}{[H^{-1}]_{jj}} H^{-1} e_j \, . \tag{2.285}$$

This result is then utilized to establish an estimate of the saliency for weight j

$$\delta V^{(j)} = V_N(\hat{\theta}^{(j)}, Z^N) - V_N(\hat{\theta}, Z^N)$$

$$\simeq \left(\hat{\theta}^{(j)} - \hat{\theta} \right)^T \left(\frac{1}{N} \sum_{t=1}^{N} \psi(t, \hat{\theta}) \epsilon(t, \hat{\theta}) \right)$$

$$+ \frac{1}{2} \left(\hat{\theta}^{(j)} - \hat{\theta} \right)^T R \left(\hat{\theta}^{(j)} - \hat{\theta} \right) \, . \tag{2.286}$$

Since $0 = W_N'(\hat{\theta}, Z^N) = V_N'(\hat{\theta}, Z^N) + \frac{1}{N} D \hat{\theta}$, the expression can be simplified to

$$\delta V^{(j)} = -\left(\hat{\theta}^{(j)} - \hat{\theta} \right)^T \frac{1}{N} D \hat{\theta} + \frac{1}{2} \left(\hat{\theta}^{(j)} - \hat{\theta} \right)^T R \left(\hat{\theta}^{(j)} - \hat{\theta} \right) \, . \tag{2.287}$$

By finally inserting (2.285) in (2.287), the following expression is obtained for the saliency

$$\delta V^{(j)} \simeq \frac{\hat{\theta}_j}{N[H^{-1}]_{jj}} e_j^T H^{-1} D \hat{\theta} + \frac{1}{2} \left(\frac{\hat{\theta}_j}{[H^{-1}]_{jj}} \right)^2 e_j^T H^{-1} R H^{-1} e_j$$

$$= \frac{\hat{\theta}_j}{N[H^{-1}]_{jj}} e_j^T H^{-1} D \hat{\theta} + \frac{1}{2} \left(\frac{\hat{\theta}_j}{[H^{-1}]_{jj}} \right)^2 e_j^T H^{-1} e_j$$

$$- \frac{1}{2N} \left(\frac{\hat{\theta}_j}{[H^{-1}]_{jj}} \right)^2 e_j^T H^{-1} D H^{-1} e_j$$

$$= \frac{1}{N} \frac{\hat{\theta}_j}{[H^{-1}]_{jj}} e_j^T H^{-1} D \hat{\theta} - \frac{1}{2N} \left(\frac{\hat{\theta}_j}{[H^{-1}]_{jj}} \right)^2 [H^{-1} D H^{-1}]_{jj}$$

$$+ \frac{1}{2} \frac{\hat{\theta}_j^2}{[H^{-1}]_{jj}} \, . \tag{2.288}$$

For completeness it should be pointed out that without weight decay the first two terms on the right-hand side will be zero, and the expression reduces to the one known from Hassibi and Stork (1993):

$$\delta V^{(j)} = \frac{1}{2} \frac{\hat{\theta}_j^2}{[R^{-1}]_{jj}} \, . \tag{2.289}$$

Compare also with (2.271).

It has been discussed previously that the Hessian of the unregularized criterion, R, generally is ill-conditioned when the network is overparametrized. To
overcome this problem Hassibi and Stork (1993) proposed that the inverse
Hessian should be approximated with the recursion used in the recursive
Gauss-Newton algorithm (see Section 2.4.1). In the regularized case (i.e.,
(2.288)) inversion of the Hessian is not a problem if only the weight decay is
sufficiently large. In general it is therefore recommended to use some weight
decay. Practical experience also indicates that weight decay improves the
robustness of the OBS-procedure.

In the regularized case, the generalization-based saliencies in (2.277) are
needed; in addition to the saliencies given by (2.288) an estimate of the
effective number of weights in the reduced architecture is thus required. Let
$\tilde{R}$ and $\tilde{H}$ specify R and H with row j and column j removed. If the second-
order expansion of the criterion is utilized once more, the effective number
of weights in the reduced network can then be approximated by

$$p_1^{(j)} = \mathbf{tr} \left\{ \tilde{R}\tilde{H}^{-1}\tilde{R}\tilde{H}^{-1} \right\} . \qquad (2.290)$$

As it turns out, it is not necessary to invert the full matrix $\tilde{H}$. Pedersen and
Hansen (1995) point out that the inverse can be found by utilization of a
well-known result about inversion of partitioned matrices:

$$A = \begin{bmatrix} A_{11} & A_{12} \\ A_{21} & A_{22} \end{bmatrix}$$

$$\Updownarrow$$

$$A^{-1} = \begin{bmatrix} S_{11}^{-1} & -S_{11}^{-1}A_{12}A_{22}^{-1} \\ -A_{22}^{-1}S_{21}S_{11}^{-1} & A_{22}^{-1} + A_{22}^{-1}A_{21}S_{11}^{-1}A_{12}A_{22}^{-1} \end{bmatrix} . \qquad (2.291)$$

The matrix S_{11} denotes the *Schur complement* of A_{11} and is defined by

$$S_{11} = A_{11} - A_{21}A_{22}^{-1}A_{12} . \qquad (2.292)$$

Without loss of generality it can be assumed that weight j is placed last in
the parameter vector. Partitioning the inverse Hessian as

$$H^{-1} = \begin{bmatrix} Q & q_j \\ q_j^T & q_{jj} \end{bmatrix} \qquad (2.293)$$

the inverse Hessian of the reduced network can be expressed as the Schur
complement of Q

$$\tilde{H}^{-1} = Q - \frac{1}{q_{jj}}q_j q_j^T . \qquad (2.294)$$

In the general case q_j denotes the jth column of H^{-1}, q_{jj} the jth diagonal element and Q is obtained by removing the jth row and the jth column from H^{-1}.

When the saliencies have been compared and the weight corresponding to the maximum decrease in FPE has been eliminated, the procedure is repeated. There has been some discussion on whether or not the network should be retrained at this point. The OBS does not only remove weights but with (2.285) it also suggests new values for the remaining weights. This feature is sometimes referred to as *re-estimation within the second-order approximation* (Larsen, 1993). In Hassibi and Stork (1993) it is claimed that this re-estimation is adequate and that there is no reason for retraining the network until the pruning session has been completed. This statement reflects a somewhat exaggerated confidence in the series expansion of the criterion. Based on practical experience it is strongly recommended that the network is retrained with suitable intervals. The safest strategy is of course a retraining after each weight elimination, but longer intervals may also work. See algorithm in Table 2.4.

Table 2.4. Pruning algorithm.

```
prune 5% of the remaining weights
retrain
prune 5% of the remaining weights
retrain
etc.
```

Alternatively, one may consider a strategy based on evaluation of the gradient of the criterion for deciding when it is necessary to retrain. As long as the gradient (in terms of some norm) remains sufficiently close to zero, retraining is not necessary. When the threshold is exceeded retraining is performed.

In principle, pruning should be continued for as long as it leads to a decrease in the FPE estimate. However, reality demands a more pragmatic view on the termination. The retraining will cause fluctuations in the FPE estimate, and the pruning session should therefore not be stopped until it seems to be a clear trend that FPE is in fact increasing. Hence, too many weights are pruned. It is thus necessary to store the architectures for a number of steps back to make it possible to re-establish the optimal architecture when it has been pointed out. To monitor how the pruning session progresses, and as an additional indicator for when to stop, it is useful to continuously evaluate the test error as well.

One final issue needs to be dealt with before it is possible to outline a complete OBS-procedure. The definition of saliencies is not appropriate when elimination of a particular weight will leave a hidden unit "floating" in the sense that it is only connected to either the inputs or the output. It is in this case necessary to investigate if the entire unit should be eliminated. Consequently, a saliency must somehow be associated with the unit rather than with the weights leading to and from it. For a two-layer network one might consider the following approach:

- In the beginning only input-to-hidden layer weights are eliminated.

- When a hidden unit has only one weight leading to it, the saliency for the entire unit is determined.

- If this unit saliency is smaller than any other saliency, the unit is eliminated.

Calculating an appropriate saliency for the concerned hidden unit is not as straightforward as one might think. It is not correct to calculate the saliency as the change in FPE obtained by fixing both weights connected to the unit to zero. The problem is that when one weight is fixed to zero the identifiability of the opposite weight is lost. Hence, one might as well calculate the saliency by fixing one weight to zero and the opposite to five or some other arbitrary value. If retraining is not employed to determine the saliency, one must rely on a suboptimal strategy. Different such strategies are possible but to compete with retraining it is important that it is simple.

Despite the fact that fixing both weights to zero is incorrect, it seems to work well in practice. For simplicity it is therefore recommended to use this solution. It is straightforward to extend the different expressions to cope with this:
Let the set $J = \{j_1, j_2\}$ contain the locations of the two weights in the parameter vector and let E_J be a matrix with the j_1th and j_2th unit vectors as its columns. Define the unit saliency by $\delta V^{(J)}$, the Lagrange multipliers by λ_J, and the two weights by $\hat{\theta}_J$. With these definitions it is easily verified that (2.284), (2.288) and (2.285) must be modified to

$$\lambda_J = \left[E_J^T H^{-1} E_J \right]^{-1} \hat{\theta}_J \tag{2.295}$$

$$\delta V^{(j)} \simeq \frac{1}{N} \lambda_J^T E_J^T H^{-1} \left(D \left[\hat{\theta} - \frac{1}{2} H^{-1} E_J \lambda_J \right] + \frac{1}{2} E_J \lambda_J \right) \tag{2.296}$$

$$\hat{\theta}^{(J)} = \hat{\theta} - H^{-1} E_J \lambda_J . \tag{2.297}$$

Furthermore, (2.294) is substituted by,

$$\tilde{H}^{-1} = Q - q_J \left[q_{JJ} \right]^{-1} q_J^T . \tag{2.298}$$

Table 2.5. Complete procedure for pruning by OBS.

1: Select a "sufficiently" large model structure and train the network with a small weight decay.
2: Compute the Gauss-Newton Hessian and invert it.
3: Evaluate the FPE estimate and evaluate the test error if a test set is available. If it is believed that both quantities have passed their respective minimum, go to *Step 7.*
4: Compute the saliency for each weight in the input-to-hidden layer by (2.288) and the effective number of parameters in the reduced network by (2.290). Combine the two into the generalization-based saliencies by (2.277). In case only one weight is leading to a hidden unit, its saliency is substituted by the "unit saliency."
5: Search for the smallest generalization-based saliency and eliminate the concerned weight(s) from the parameter vector. Determine the remaining weights by the projection (2.285) or (2.297) if an entire unit is removed).
6: Retrain the network, store the weights, and go to *Step 2.*
7: Re-establish the architecture that was optimal as assessed by the test error and FPE estimate, and retrain and retrain the network without weight decay.

A complete procedure for pruning with the OBS algorithm is shown in Table 2.5.

If the Levenberg-Marquardt method is used for retraining the network in *Step 6*, 10–50 iterations are usually sufficient.

It is emphasized that the OBS procedure is not always equally successful in practice. Not even when the training set is large and an appropriate initial network has been found. The two main problems with OBS are:

- The second-order expansion of the criterion is valid only in the proximity of the minimum. When OBS suggests how a weight should be set to zero and the remaining weights re-estimated, it may violate the validity of the series expansion.

- The fundamental assumption governing OBS is that the optimal architecture can be achieved by stripping off weights one at a time in accordance with what the (generalization-based) saliencies suggest is the optimal in that particular step. One cannot expect that this strategy will in fact reach the optimal network.

A possible remedy for the first problem is to compare the estimated saliency to an evaluation of the network with the re-estimated weights. If the discrepancy exceeds, say, 10%, retraining should be used to determine the exact saliency. One might also consider combining OBS with an advanced Levenberg-Marquardt method that estimates the trust region (see Section 2.4.1). It can then be checked if OBS suggests a modification of the weights that are outside the trust region. If so, the network is retrained to determine the saliency more accurately.

The second problem is more difficult to overcome. In fact, some argue that for this very reason it is simply better to rely on regularization instead of pruning and concentrate on methods for finding the optimal setting of the weight decay parameter.

It can be argued that the problem justifies a more relaxed attitude towards the chosen pruning algorithm. Thus, one should be quite free to consider different short-cuts that will reduce the amount of computations. One such short-cut is described in Hansen et al. (1994). It is proposed to prune the weights according to (training-error-based) saliencies just as in the unregularized case. In practice the strategy appears to perform equivalent to the generalization-based method, but one avoids having to calculate the effective number of weights.

To illustrate what a pruned network might look like, Figure 2.24 compares a fully connected NNARX(2,2,1) model structure with 91 weights to a model structure obtained by pruning the network until it contained only 19 weights. An example of OBS used in modelling of a real time series can be found in Chapter 4.

Optimal Brain Damage (OBD). Optimal Brain Damage presents an earlier and somewhat simpler approach to network pruning (Le Cun et al., 1990). The scheme is based on the assumption that all off-diagonal elements in the Hessian matrix can be neglected, which is similar to assuming that all the weights are mutually uncoupled. Unlike OBS, OBD only provides guidelines for pruning and not for how to re-estimate the remaining weights. This fact explains the motive behind the name convention for the two schemes: when brain damage removes a weight, it leaves the remaining weights untouched; brain surgeon removes a weight and modifies the remaining weights.

The jth element in the vector $\delta\theta$ is the only element that is different from zero. Hence, it is easily verified that the saliency of weight j is given by

$$\delta V^{(j)} = \left(\frac{D_{jj}}{N} + \frac{1}{2} R_{jj} \right) \hat{\theta}_j^2 \,, \qquad (2.299)$$

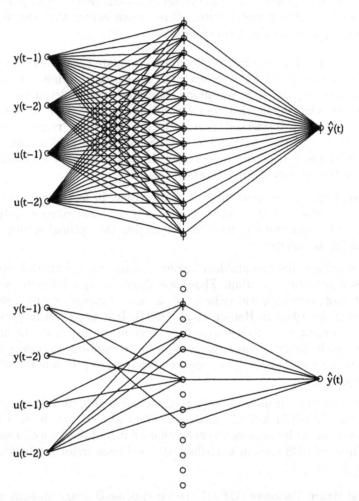

Figure 2.24. The initial, fully connected architecture (top) and the network pruned to only 19 weights (bottom). A vertical line through a unit symbolizes a bias (and is counted as a weight).

and the generalization-based saliency of weight j by

$$\delta V_M^{(j)} = \delta V_M^{(j)} - \frac{2}{N} \left(\frac{R_{jj}}{R_{jj} + \frac{D_{jj}}{N}} \right)^2 V_N(\hat{\theta}, Z^N) \,. \qquad (2.300)$$

Since the coupling of weights is ignored there is no re-estimation within the second-order approximation to carry out between weight eliminations. This has the effect that floating hidden units will not cause problems as they do in OBS. Additionally, elimination of more than one weight at a time is "allowed."

The amount of computations required by OBD is substantially smaller than for OBS and although it is based on an extremely crude approximation, it often turns out to perform similarly to OBS.

Both OBD and OBS can be implemented in a highly automated fashion. The stage that demands most user interaction is the search for an initial model whose structure (in principle) is large enough to describe the system. This implies that different model orders, different types of regressors, and a different number of hidden units have to be tried out. Obviously, the amount of data available will impose an upper limit to the number of weights that one is allowed to include in the initial model structure. If it is not possible to describe the system satisfactorily with the available training data, it may be necessary to redo the experiment.

2.6.3 Redoing the Experiment

Sometimes it is necessary to experiment further with the system to generate additional data for inclusion in the training set. If lack of data is the problem this will often be revealed when comparing the actual outputs in the test set to the predictions provided by the network model. If a particular regime is not adequately represented in the training data, there will often be systematic discrepancies between actual outputs and predictions when the network is applied in this regime. The prediction intervals proposed in Section 2.5 can be an useful tool for pointing out such regimes where more data should be collected.

2.6.4 Section Summary

In this section it was discussed how to iterate in the identification procedure. That is, in which ways one must expect to go back to previous stages in the procedure in order to improve the model. Particular emphasis was put on automatic methods for model structure selection; the so-called pruning algorithms.

Retraining the network. As the criterion will have a number of local minima, training will not necessarily lead to the global minimum the first time around. In fact, the found minimum can be quite far in value from the global minimum. The network should therefore be trained a couple of times, starting with different initializations of the weights. This is typically done right away; it is not necessary to validate each network first. Thus, it is not really a matter of going backwards in the procedure.

Another issue mentioned was that one must expect to alter the criterion by tuning the weight decay parameter(s) in the regularization term. This is done in order to improve generalization.

Selecting the optimal model structure. Two different pruning algorithms were described: *Optimal Brain Surgeon (OBS)* and *Optimal Brain Damage (OBD)*. In both cases the principle is to initially start out with a relatively large network architecture and then successively prune the network branches (weights) of one at a time until the optimal architecture has been found. As a stopping criterion one should use one or more of the estimates of average generalization error described in Section 2.5, e.g., FPE, (LU)LOO, or the test error.

The difference between the two strategies for pruning is that while OBD only prunes weights, OBS also recommends how to change the value of the remaining weights in the network. For the best result it was recommended to retrain the network each time a weight had been eliminated. However, one can often get away with pruning a few weights between each retraining.

Continuing the experiment. Often the final model is simply not good enough despite that one has made a large effort in extracting as much information as possible from the data. It was mentioned that the problem in this case laid somewhere else; namely in the available data. It is quite common that one has to acquire additional data after having validated the model and located the regimes where its performance is unacceptably poor.

2.7 Recapitulation of System Identification

A procedure for system identification was proposed in the beginning of this chapter and the subsequent sections provided detailed descriptions on how each stage in the procedure could be realized. To conclude the chapter it is thus appropriate to gather up the threads by briefly outlining a generic working procedure for system identification with neural networks. One should of course be aware that such a procedure must always be squeezed a little here and there to conform to the application under consideration.

It is assumed that it has been found necessary to try a neural network approach. This can be motivated by physical insight; because a linear approach failed; or because nonlinearity tests, such as those discussed in Section 2.3, indicated that a nonlinear model is necessary.

Overall aspects

- The chapter dealt with several advanced techniques for extracting as much information as possible from the observed data. When the set of training data is not very large, most of these techniques become vital and so does any physical insight that one might possess. However, in many cases one has access to a very large set of data in which case issues like regularization and pruning essentially are irrelevant. This is often the case for applications related to control as here it is relatively easy to generate data.

- One should always have a bias towards simple solutions. Care must be taken to avoid an uncritical pursuit for the model providing the best possible (one-step ahead) predictions if this is not the actual objective. For instance, in many applications it is more important that the dimension of the regression vector is low, that the network contains few weights and that the predictor is stable and a reasonably smooth function of the regressors.

- A neural network is not a *statistician-in-a-box*. It does not perform miracles and a sound understanding of basic statistics is a natural requirement if one is to have success.

- Adaptive systems are sometimes called a "fiddler's paradise." It is important that the time aspect is also taken into consideration when neural networks are used in system identification. In many cases it is better to acquire more data or to make some kind of intelligent pre-processing of the data than to play around with numerous choices of regressors, network architectures, and regularization schemes.

Experiment

It is crucial that one is fully aware of the curse of dimensionality when the experiment is conducted. If the system is nonlinear, it is of vital importance that the entire operating range is represented in the data set; frequency-wise as well as amplitude-wise. Because of this and because neural networks typically contain many adjustable parameters (the weights), one should make sure to collect as much data as possible.

When a set of data has been acquired one should start with a visual inspection of it to determine if it needs additional filtering, or there are outliers that must be removed. It may occur that certain information is extremely well represented in the data. Unless one in fact wish to weight this information higher at the expense of the information in the remaining part of the data set, the redundancy should be removed. Finally the input and output sequences should be scaled to zero mean and variance one.

The data set is then divided into a training and a test set, respectively. A common choice is half-and-half, but the amount of data may motivate other divisions. Particularly when the data set is small, it is common to include a larger fraction of the data in the training set. In fact one can learn valuable insight from repeating the procedure for different *split ratios*. The graph of the estimated test error as function of the size of the training data set is called the *learning curve* and expresses the effectiveness of the learning procedure. See, e.g., Gorodkin et al. (1993).

Model structure selection

Selection of an appropriate model structure is difficult due to the many degrees of freedom. It is not possible to determine *the* optimal model structure, so instead one should choose a strategy that is reasonably effortless to follow and that finds a structure which is sufficiently close to the optimal. The choice of strategy depends heavily on the circumstances. Three typical cases are:

The amount of data is small: If the amount of data is small it is not possible to select a model structure that is "large enough". Thus, the bias error will typically contribute substantially to the average generalization error. It is an unfortunate situation to be in: generalization error will always be high, it is hard to spare any data for a test set, the foundation for the FPE estimate (an in turn for the pruning algorithms) is lost when the undermodelling is significant. Model validation and comparison is difficult.

It is recommended to consider only NNARX model structures based on fully connected network architectures. Model structures based on recurrent network architectures should be avoided since the lack of excitation in certain regimes might lead to problems with instability of the predictor. As a rule of thumb, the model structure should be chosen so that the ratio between the number of training data and the weights in the network, N/p, is between 2 and 7. A small ratio when the signal-to-noise ratio is high and large ratio when the signal-to-noise ratio is poor. The network should be trained without a regularization term as this will only introduce an additional bias error.

The amount of data is large: When the amount of data is large, the effect of regularization and pruning will be insignificant. For this reason it is only necessary to consider fully connected networks trained without a regularization term. If the noise level is high it might be worth while spending some time on trying the more advanced model structures like NNARMAX and NNSSIF. The network architecture is determined by gradually increasing the number of hidden units until the test error levels off.

The amount of data is medium-sized: This is an important case since it is important that a suitable compromise between bias and variance error is found. Two techniques for dealing with the issue have been discussed: *regularization by weight decay* and *pruning*. Either way, one should first determine an initial model structure based on a fully connected network which is so large that the bias error can be ignored. The number of hidden units in this network is found by gradually increasing the architecture with one hidden unit at a time. For each number of hidden units, the network is trained to the minimum 5–7 times, starting with a different initialization of the weights each time. When it appears to be a clear trend that the test error increases, one should stop and pick a model structure that is slightly bigger than the optimal one (2–5 extra hidden units). At this point one can choose one of two strategies:

- An accurate determination of the optimal weight decay. For simplicity one should consider only a single weight decay parameter.

- Prune the network. OBS is in general preferred over OBD since it is derived on a more sound statistical foundation.

It is often difficult to know in advance in which group a particular identification problem belongs. Whether the data set is small or large must somehow be measured relative to the complexity of the system to be identified. To gain a feel for the character of the present problem, it might be necessary to "iterate" a few times in the identification procedure.

The NNARX model structure should always be the first choice of structure (or perhaps an NNFIR structure if the system is stable and well-damped). The advantage of an NNARX model is that none of the regressors depend on past outputs of the model, which ensures that the predictor remains stable. Not only does this facilitate training, it will in general result in more robust models. Moreover, if an NNARX structure turns out to perform miserably, it is not likely that one will succeed with the more advanced model structures either.

It is advantageous if the decision of the number of past inputs and outputs to be included in the regression vector can be motivated by physical insight. However, when the knowledge about the system is limited one may instead consider the method based on Lipschitz quotients, which was explored in Section 2.2.5. If the system is deterministic (or the signal-to-noise ratio is high) this method can often provide a reasonable estimate of the order. The Lipschitz quotients can also be used to detect a possible time-delay, but a more reliable approach is here to study the cross-correlation between input and output.

Estimate a model

This encompasses two tasks:

- Select a criterion specifying how the weights should be determined from data.

- Choose an iterative search method for minimizing the criterion (a *training algorithm*).

The present chapter considered a mean square error criterion with and without a regularization term. The choice of criterion is thus a matter of selecting a suitable weight decay. As was mentioned before, regularization is primarily relevant if the set of training data is medium-sized. However, in preliminary investigations, when determining an initial network architecture that is large enough to describe the system, regularization should not be used at all.

In order to train the network it is generally recommended to apply the Levenberg-Marquardt method. This method is derived especially for mean square error type criteria. It provides a fast convergence, it is robust, simple to implement, and it is not necessary for the user to initialize any strange design parameters.

The advanced developer may consider a scheme where the Quasi-Newton method is used in the final iterations. If the Levenberg-Marquardt method has trained the network to a point that is reasonably close to the minimum (in that the Hessian is close to the Hessian evaluated in the minimum), a Quasi-Newton method will typically speed up convergence.

Validation

It is always desirable that the trained neural network model is (also) validated on a set of data that was not used for training the network. Exactly how the validation is carried out is of course highly dependent on the intended use of the model.

Different means of validation were discussed in Section 2.5:

- Correlation tests to investigate if the residuals are white and independent of past information.

- Visual inspection of *one*-step and k-step ahead predictions. Perhaps also estimation of prediction intervals.

- Estimation of the average generalization error.

If a fully connected network model trained without regularization is able to pass the correlation tests and the predictions look reasonably good, the model can essentially be accepted right away. However, if the ratio between number of training data and weights in the network, N/p, is smaller than 10–15, it can be worth spending some extra time on adjusting the weight decay and/or prune the network.

Going backwards in the procedure

Retrain the network: Because the network usually has more than one local minimum it is recommended to always train the network 5–7 times with different initializations of the weights.

Determine the optimal weight decay: If one has decided to rely on regularization for finding a suitable trade-off between bias and variance errors, the trail-and-error approach, discussed in Section 2.4.2, is recommended. A combination of FPE, (LU)LOO, and test error can be used for pointing out the optimal weight decay parameter. When available, the test error should be the primary indicator.

Determine the optimal model structure by pruning: Determine a weight decay that is somewhat smaller than the optimal. Prune one weight at a time and let it be followed by a retraining. Retrain the network with 50ʹ Levenberg-Marquardt iterations. Use, as a stopping criterion, a combination of FPE, (LU)LOO, and test error.

Model accepted

When the final model has been found, the weights should be rescaled, so that the model can be applied directly to unscaled data.

3. Control with Neural Networks

The previous chapter dealt with principles for using neural networks in identification of nonlinear systems. Although this issue has relevance to many different types of applications, in this book it is regarded primarily as one of the steps required in the development of controllers for unknown nonlinear systems. Control of nonlinear systems is a major application area for neural networks. The control design problem, which is the subject under investigation in this chapter, will be approached in two ways: *direct design methods* and *indirect design methods*. "Direct design" means that a neural network directly implements the controller. Therefore, a network must be trained as the controller according to some kind of relevant criterion. The indirect methods represent a more conventional approach, where the design is based on a neural network model of the system to be controlled. In this case the controller is not itself a neural network. A number of designs belonging to each of the two categories will be presented and their characteristics will be discussed. Much attention will be drawn to the implementation aspects.

3.1 Introduction to Neural-Network-based Control

To *control* a system is to make it behave in a desired manner. How to express this "desired behavior" depends primarily on the task to be solved, but the dynamics of the system, the actuators, the measurement equipment, the available computational power, etc., influence the formulation of the desired behavior as well. Although the desired behavior obviously is very dependent of the application, the need to rephrase it in mathematical terms suited for practical design of control systems seriously limits the means of expression. At the higher level it is customary to distinguish between two basic types of problems:

Regulation problems. The fundamental desired behavior is to keep the output of the system at a constant level regardless of the disturbances acting on the system (e.g., controlling the temperature in a room).

Servo problems. The fundamental desired behavior is to make the output follow a reference trajectory closely (e.g., controlling a robot).

As for most other textbooks on control, it will probably appear that this book has a bias towards considering servo problems. However, disturbance rejection is also considered in several sections throughout the chapter.

On a secondary level the desired behavior must be phrased in a manner that more directly influence the actual choice of design method. In this work two basic formulations of the desired behavior are accepted:

- The closed-loop system, consisting of controller and system to be controlled, should follow a prescribed transfer function model. This class of design methods comprises well-known strategies like pole placement and model-reference controllers.

- Express the desired behavior in terms of a quadratic criterion and derive the controller as the minimizer of this criterion. Examples are minimum variance, predictive, and optimal control.

The former class of designs is typically based upon using feedback to obtain a cancellation of the nonlinearities in the system. This way, the closed-loop system from reference to output will behave linearly, but the behavior of the controller itself is typically hard to specify simultaneously. In the latter case, controller as well as closed-loop system will be nonlinear, but tuning can typically be done by quite intuitive means.

Often it is preferable to formulate the behavior in terms of time domain characteristics: keep the output of the system below a certain value at all times, design the controller to give an overshoot of 5% or less, the rise-time should be 8 seconds or less, the control signal should be smooth, etc. These types of desired behaviors can often be handled under linear conditions and to some extent even in the stochastic case. Unfortunately, they are in general hard to satisfy for unknown nonlinear systems. Often one is forced to apply one of the two ways of specifying the behavior mentioned above and then tune the controller by simulation to give the desired overshoot, rise-time, etc. The simulation must naturally be very extensive, since it is necessary to check the behavior of the controlled system over the entire range of operation.

A universal requirement to the behavior of the closed-loop system is that it should be stable. Unfortunately, the stability issue is most often a complicated matter for nonlinear systems. For example, it may occur that a nonlinear system locally is unstable in certain regimes while it is stable in others. Whether this will be a problem in practice therefore depends on how the system is operated (the reference trajectory). Analysis of nonlinear dynamics and design of globally stabilizing controllers for nonlinear systems are topics, which are addressed in many textbooks. See for example Nijmeijer and van der Schaft (1990) and Slotine and Li (1991). Regardless of the amount of work done in the field, this presentation will not embark on a treatment of stability issues. The requirements for designing stable control systems are usually quite strict.

It is limited to certain classes of nonlinear systems and it is assumed that an accurate model of the system exists or that, at least, the inaccuracies are well-defined and their magnitude known. Furthermore, a conservative controller is a frequent result of the design procedure. These limitations are in contradiction with the generic and pragmatic approach attempted here and for this reason an in-depth treatment of stability issues will not be given.

The location of the zeros of a system is important in relation to the feasibility of several linear controllers. Since nonlinear systems cannot be described in terms of transfer functions, it is not possible to use the phrase "zeros" here (except in a local sense). As for the question of stability, to which this issue is related, the absence of distinct zeros impedes nonlinear control design further. Although it may seem as an abuse of terminology, the phrase "zeros" will occasionally be used as a somewhat fuzzy extension of the linear case. Section 3.7 will provide a justification for this. In addition, the phrases: "an unstable inverse" and "a lightly damped inverse" will often be used as counterparts to the linear case where there are zeros outside the unit circle and zeros inside, but close to the unit circle, respectively.

It is often seen that neural networks are proposed as a tool for adaptive control of nonlinear systems with time-varying dynamics. Although the idea sounds intriguing, one must be sceptical regarding the practical usefulness of such schemes. Neural networks are too complex to be considered in relation to control of most time-varying systems: an adaptive controller based on a neural network is simply too flexible a strategy to become successful in practice. Only in certain simple cases, for example when the time-variations are relatively slow, there is a chance that an adaptive control scheme can be brought to work. Regardless of these difficulties the present chapter includes a few controller designs founded on on-line adaption of a neural network. However, these schemes are primarily intended to be applied to time-invariant systems. They should only be used until the network has converged. When the behavior of the closed-loop system is satisfactory, the adaption should be stopped.

The chapter covers two fundamentally different approaches to neural-network-based control:

A direct control system design. "Direct" means that the controller is a neural network. A neural network controller is often advantageous when the real-time platform available prohibits complicated solutions. The implementation is simple while the design and tuning are difficult implying a retraining of the network every time a design parameter is modified. Often this training has to be performed according to an on-line scheme. With a few exceptions this class of designs is *model-based* in the sense that a model of the system is required in order to design the controller. The concepts highlighted in this chapter (Section 3.2 to Section 3.6) include:

- Direct inverse control (including training of inverse models).

- Internal model control.

- Feedback linearization.

- Feedforward with inverse models.

- Optimal control.

An indirect control system design. This class of designs is always model-based. The idea is to use a neural network to model the system to be controlled. This model is then employed in a more "conventional" controller design. The model is typically trained in advance, but the controller is designed on-line. As it will appear, the indirect design is very flexibile; thus, it is the most appropriate for the majority of common control problems. The indirect design methods covered in Section 3.7 and Section 3.8 include:

- Approximate pole placement, minimum variance, and predictive control.

- Nonlinear predictive control.

Except for a few minor deviations, each section of this chapter has been organized in a similar fashion:

1. A control design is presented.

2. The respective control law is derived and its properties are discussed.

3. The performance is studied in a small simulation example.

4. The most important features of the design are briefly summarized.

3.1.1 The Benchmark System

To illustrate the characteristics of the different controllers, the same simple system will be considered throughout the chapter. In selecting this system it was essential that it would be possible to control it with all the methods described in the chapter. An open-loop stable system with smooth nonlinearities for which also the inverse is unstable has therefore been chosen Soloway and Haley (1996). The system is given by

$$\ddot{y}(t) + \dot{y}(t) + y(t) + y^3(t) = u(t) \,. \tag{3.1}$$

Being this simple, naturally it cannot illustrate all the important features of each controller. However, it was assumed that the chapter would be more readable if the same system was used everywhere. After all, Chapter 4 contains more challenging applications.

To explore the nonlinearities in the system, two simple open-loop experiments were carried out. First, three square waves of different amplitudes were applied to the system (Figure 3.1) and secondly, three sine waves of different amplitudes (Figure 3.2) were applied. The sampling frequency was selected to $f_s = 5$ Hz.

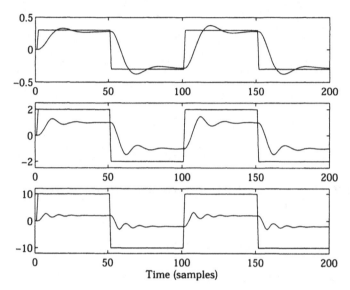

Figure 3.1. Open-loop simulation of benchmark system with three different square waves applied as input.

Looking at the two figures it is quite clear that the system is nonlinear. In particular as the magnitude of the applied input signal is increased. Figure 3.2 shows that the response is not even close at being sinusoidal when the amplitude of the input is 10. When applying the squares (Figure 3.1) it is evident that the system becomes less damped as the amplitude is increased. Both figures show that the homogeneity condition is not satisfied

$$y(t) = g(\alpha u(t)) \neq \alpha g(u(t)) . \tag{3.2}$$

3.2 Direct Inverse Control

When neural networks originally were proposed for controlling unknown nonlinear systems, one of the first methods being reported was on training a network to act as the inverse of the system and use this as a controller. Explained in brief, the basic principle is as follows:

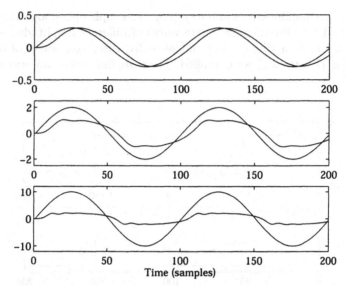

Figure 3.2. Open-loop simulation of benchmark system with three different sine waves applied as input.

Assuming that the system to be controlled can be described by

$$y(t+1) = g[y(t), \ \dots \ , y(t-n+1), \ u(t), \ \dots \ , u(t-m)] \qquad (3.3)$$

The desired network is then the one that isolates the most recent control input, $u(t)$,

$$\hat{u}(t) = \hat{g}^{-1}[y(t+1), y(t), \ \dots \ , y(t-n+1) \ , u(t), \ \dots \ , u(t-m)] \ . \qquad (3.4)$$

Assuming such a network has somehow been obtained, it can be used for controlling the system by substituting the output at time $t+1$ by the desired output, the reference, $r(t+1)$. If the network represents the exact inverse, the control input produced by it will thus drive the system output at time $t+1$ to $r(t+1)$. The principle is illustrated in Figure 3.3 for a first-order system.

This section will describe two techniques for establishing the inverse model: an off-line method known as *general training* and an on-line method called *specialized training*. Properties and limitations of inverse model controllers will be discussed and so will their relationship to well-known linear controllers. The benchmark system introduced in Section 3.1.1 will be used for illustrating several of the issues addressed.

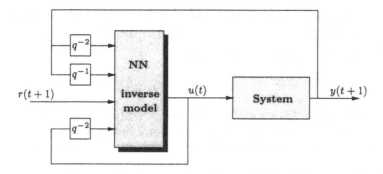

Figure 3.3. Direct inverse control.

3.2.1 General Training

The most straightforward way of training a network as the inverse of a system is to approach the problem as a system identification problem analogous to the one considered in Chapter 2: an experiment is performed, a network architecture is selected, and the network is trained off-line. The difference from system identification lies in the choice of regressors and network output. They are now selected as shown in (3.4). The network is then trained to minimize the criterion

$$J(\theta, Z^N) = \frac{1}{2N} \sum_{t=1}^{N} [u(t) - \hat{u}(t|\theta)]^2 \ . \tag{3.5}$$

This strategy is referred to as general training. The network can be trained by using any of the training methods presented in Section 2.4. Extension to multivariable systems is straightforward.

An appealing feature of general training is that fundamentally it is not a model-based design method. This should be understood in the sense that the controller is inferred directly from a set of data without requiring an actual model of the system.

One should not be confused by the assumption that the reference is known one step in advance (see Figure 3.3). This was just used as a pedagogical notation. The closed-loop transfer function from reference to output of the system is in fact

$$H(q^{-1}) = q^{-1} \ . \tag{3.6}$$

One can interpret this as if the controller is linearizing the system, resulting in a dead-beat controller: the system output will follow the reference signal exactly except for a delay of one sampling period. If the system has a time delay exceeding one, using inverse models for control becomes slightly more complicated. A dead-beat controller is still obtained but now the closed-loop transfer function becomes $H(q^{-1}) = q^{-d}$, with d being the time delay. The principle for handling time delays is outlined in the following.

Assume that the system is governed by

$$y(t + d) = g[y(t + d - 1), \ldots, y(t + d - n),\ u(t), \ldots, u(t - m)]\ . \qquad (3.7)$$

For convenience past inputs and outputs are used directly as arguments to the function g instead of the regression vector notation employed in Chapter 2. The regression vector notation will not be used again until Section 3.7.

Once again, a network is trained as the inverse model

$$\hat{u}(t) = \hat{g}^{-1}[\ y(t + d), y(t + d - 1), \ldots, y(t), \ldots, y(t + d - n)\ ,$$
$$u(t - 1), \ldots, u(t - m)\ ]\ . \qquad (3.8)$$

Analogous to the case considered before, $y(t+d)$ is substituted for the desired output at time $t + d$. This leaves $d - 1$ unknown quantities:

$$\{y(t + 1), \ldots, y(t + d - 1)\}\ .$$

A solution to this problem is to insert predictions of the unknown outputs in (3.8), which implies that one or more networks are trained to provide the necessary predictions. Alternatively the predictor can be "incorporated" in the inverse model directly. Assume for example that the time delay $d = 2$. In this case there will be one unknown quantity, namely $y(t+1)$. The prediction of this takes the form

$$y(t + 1) \simeq \hat{y}(t + 1) = \hat{g}_1[y(t), \ldots, y(t + 1 - n),$$
$$u(t - 1), \ldots, u(t - m - 1)]\ . \qquad (3.9)$$

The inverse model is then trained by using as regressors the union of regressors from (3.8) and (3.9)

$$\hat{u}(t) = \hat{g}^{-1}[y(t + 2), y(t), \ldots, y(t + 2 - n), \ldots, y(t + 1 - n)\ ,$$
$$u(t - 1), \ldots, u(t - m),\ u(t - m - 1)]\ . \qquad (3.10)$$

When $d > 2$, it is straightforward to proceed in a similar fashion.

What is then the practical applicability of direct inverse control? Dead-beat controllers are well-known from most standard textbooks on digital control of linear systems. See for example Åström and Wittenmark (1990). In general, however, the practical relevance is considered limited due to a number of serious inconveniences. Using feedback to achieve a fast response to reference changes will typically result in a poor robustness with a high sensitivity to noise and high frequency disturbances. In addition, one will often encounter a very active control signal. If the system is linear this occurs when the zeros are situated close to the unit circle. In the nonlinear case there is not a unique set of zeros, but of course a similar phenomenon exists.

If the inverse model is unstable (corresponding to zeros outside the unit circle in the linear case), one must anticipate that the closed-loop system becomes unstable. Unfortunately, this situation occurs quite frequently in practice. In Åström et al. (1984) it is proved that discretization of linear continuous-time models under quite common circumstances can result in zeros outside the unit circle regardless that the continuous-time model has no zeros, or all zeros are in the left half plane. In fact, for a model with a pole excess of at least two, one or more zeros in the the discretized model will converge to the unit circle or even outside as the sampling frequency is increased. It must be expected that a similar behavior also can be found in discrete models of nonlinear systems. An instrument for exploration of such problems will be given in Section 3.7.

Another problem with the design arises when the system to be controlled is not one-to-one since then a unique inverse model does not exist. That is,

$$g[y(t), \ \dots \ y(t - n + 1), u_1(t), \ \dots \ , u(t - m)] =$$
$$g[y(t), \ \dots \ y(t - n + 1), u_2(t), \ \dots \ u(t - m)] \qquad (3.11)$$

might occur for two different control inputs, $u_1(t) \neq u_2(t)$. If this non-uniqueness is not reflected in the training set one can in principle yield a particular inverse which might be adequate for controlling the system. Most often, however, one will end up with a useless incorrect inverse model.

There is also a problem in relation to training the network on realistic signals. This is to be understood in the sense that in the initial experiment one should apply signals which are of a similar character as those the network is expected to produce during operation. The problem is well-known in the system identification community and the subject is frequently referred to as *identification for control*. Explained briefly the problem is that *a priori* it is unclear what realistic data means. This is not known until the inverse model has actually been trained and used as controller. Consequently, one will often benefit from an iterative design procedure where the experiment is repeated with the inverse model in the loop. A new data set is then collected and the training is repeated on this data set.

3.2.2 Direct Inverse Control of the Benchmark System

The top panel of Figure 3.4 shows the input signal applied to the benchmark system from Section 3.1.1 during the initial experiment. The response produced by the system is shown in the panel below. On this data set a network with seven *tanh* hidden units and a linear output unit was trained as the inverse model.

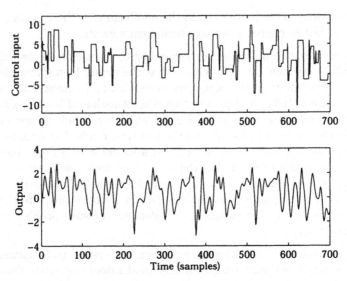

Figure 3.4. Training data set. Upper panel: control signal. Lower panel: output signal.

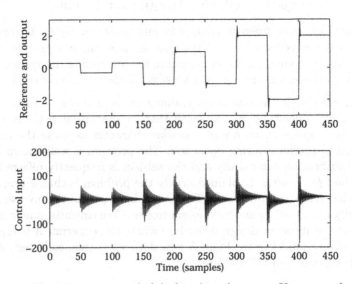

Figure 3.5. Direct inverse control of the benchmark system. Upper panel: reference and output signal. Lower panel: control signal.

The network obtained by general training was then applied as controller for the system. Figure 3.5 shows the closed-loop system's response to different step changes in the reference signal.

Apparently there is a small overshoot indicating that the inverse model is not perfect. However, apart from these inaccuracies, the inverse model on the whole manages to control the system quite well. Another effect that leaps to the eye is that the control signal appears to be extremely active and assumes very large values. This is a quite common outcome when using the present type of controller, and it has to do with the cancellation of zeros discussed above. It is easy to verify this by linearizing the system under stationary conditions. The Laplace transform of the linearized continuous model is

$$H(s) = \frac{1}{s^2 + s + 1} . \tag{3.12}$$

Discretizing this, while assuming zero-order hold on the input, leads to the following model ($fs = 5$ Hz):

$$H_0(z^{-1}) = \frac{0.0187z^{-1} + 0.0175z^{-2}}{1 - 1.7826z^{-1} + 0.8187z^{-2}} \tag{3.13}$$

Notice that the transfer function has been specified in the Z-domain (using the complex variable z) rather than in the time domain (in the delay operator q^{-1}), although only the latter has been employed previously. This is because it is more convenient when discussing poles and zeros of the transfer function. One can say that there is a pole in $z = p$, but it makes no sense to write that there is a pole in $q = p$. Occasionally, the Z-transforms will therefore occur throughout the remaining part of this chapter.

The zero is situated in $z = -0.9354$, which is close to the unit circle. One should always be cautious when cancelling negative real zeros since this will lead to rapid oscillations in response to step changes in the reference Franklin et al. (1998). Furthermore, with the zero as close to the unit circle as in the present case the control signal will also assume large values. This large, oscillating control signal is also the key in explaining the overshoot (i.e., the inaccuracy of the inverse model). When comparing the control signal to the signal applied to the system during the experiment, it is evident that they have very different shapes. Both in terms of frequency content and amplitude size. In other words, the inverse model is extrapolating in the transient periods and basically one cannot expect it to behave well in these regimes. A simple low-pass filtering of the reference by

$$H_m(q^{-1}) = \frac{0.09}{1 - 1.4q^{-1} + 0.49q^{-2}} \tag{3.14}$$

helps significantly on the model-following property as shown in Figure 3.6.

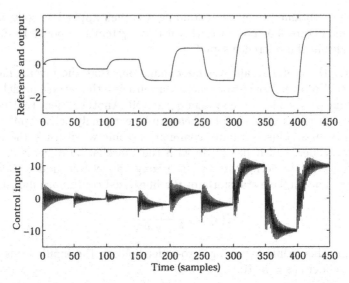

Figure 3.6. Direct inverse control after a low-pass filtering of the reference trajectory. Upper panel: reference and output signal. Lower panel: control signal.

3.2.3 Specialized Training

As discussed in Section 2.5 concerning validation of identified models, it is often difficult to quantify what exactly a "good model" means. In the context of training inverse models, which are to be used as controllers, the trained inverse model somehow ought to be validated in terms of performance of the final closed-loop system. As discussed previously, the closed-loop system should ideally exhibit a nice dead-beat behavior. This points out a serious weakness associated with the general training method: the criterion expresses the objective to minimize the discrepancy between network output and a sequence of "true" control inputs. This is not really a relevant objective. In practice it is not possible to achieve zero generalization error and consequently the trained network will have certain inaccuracies. Although these generally are reasonably small, in terms of the network output being close to the ideal control signal, there may be large deviations between the reference and the output of the system when the network is applied as controller for the system. Occasionally this is formulated as the training procedure not being *goal directed* (Psaltis et al., 1988). The goal is of course that in some sense the system output should follow the reference signal closely. It would be more desirable to minimize a criterion of the following type

$$J(\theta, Z^N) = \frac{1}{2N} \sum_{t=1}^{N} [r(t) - y(t)]^2 , \tag{3.15}$$

which clearly is goal directed. Unfortunately the minimization of this criterion is not easily carried out off-line, considering that the system output, $y(t)$, depends on the output of the inverse model, $u(t-1)$. Inspired by the recursive training algorithms explored in Chapter 2 the network might alternatively be trained to minimize

$$J_t(\theta, Z^t) = J_{t-1}(\theta, Z^{t-1}) + [r(t) - y(t)]^2 . \qquad (3.16)$$

This is an on-line approach and therefore the scheme constitutes an *adaptive controller*. Deriving a training scheme based on this criterion is not completely straightforward. A few approximations are required to make implementation possible.

By way of introduction, a recursive gradient method is considered. Assuming that J_{t-1} has already been minimized, the weights are at time t adjusted according to

$$\hat{\theta}(t) = \hat{\theta}(t-1) - \mu \frac{de^2(t)}{d\theta} , \qquad (3.17)$$

where $e(t) = r(t) - y(t)$, and

$$\frac{de^2(t)}{d\theta} = -\frac{dy(t)}{d\theta} e(t) . \qquad (3.18)$$

By application of the chain rule, the gradient $\frac{dy(t)}{d\theta}$ can be calculated by

$$
\begin{aligned}
\frac{dy(t)}{d\theta} &= \frac{\partial y(t)}{\partial u(t-1)} \frac{du(t-1)}{d\theta} \\
&= \frac{\partial y(t)}{\partial u(t-1)} \left[\frac{\partial u(t-1)}{\partial \theta} + \sum_{i=1}^{n} \frac{\partial u(t-1)}{\partial y(t-i)} \frac{dy(t-i)}{d\theta} \right. \\
&\quad \left. + \sum_{i=2}^{m} \frac{\partial u(t-1)}{\partial u(t-i)} \frac{du(t-i)}{d\theta} \right] . \qquad (3.19)
\end{aligned}
$$

It appears that the Jacobians of the system, $\frac{\partial y(t)}{\partial u(t-1)}$, are required. These are generally unknown since the system is unknown. To overcome this problem, a *forward model* of the system is therefore identified to provide estimates of the Jacobians

$$\frac{\partial y(t)}{\partial u(t-1)} \simeq \frac{\partial \hat{y}(t)}{\partial u(t-1)} . \qquad (3.20)$$

The forward model is obtained by system identification as described in Chapter 2. In principle any of the proposed model structures can be used; deterministic as well as stochastic.

The name *specialized training* and the principle behind it are usually credited to Psaltis et al. (1988). In this reference, the fact that the inverse model is a recursive network is ignored during training in the sense that the regressors's dependency on the weights in the network is ignored

$$\frac{du(t-1)}{d\theta} = \frac{\partial u(t-1)}{\partial \theta} \; . \tag{3.21}$$

The simplified training algorithm, which appears in accordance with this approximation, in some sense follows the the spirit of the so-called *recursive pseudo-linear regression methods (PLR)* Ljung (1999). These methods are quite common for recursive estimation in adaptive controllers. The algorithm is somewhat simpler to implement, the amount of computations required at each sample is reduced, and practice has shown that the training algorithm in general converges quite well anyway. A reduced computational burden is not an unimportant factor when the training has to be conducted on-line; this is probably the main motivation for making the approximation. Corresponding to the strictly positive real condition for convergence of PLR methods discussed in Ljung (1999), it must be assumed that, in a similar fashion, there are limitations on the convergence of the simplified training algorithm.

An approximation which is unavoidable in practice, is the above mentioned approximation of the Jacobians of the system. Fortunately, inaccuracies in the forward model need not have a harmful impact on the training. The Jacobian is a scalar factor and in the simplified algorithm it will therefore only modify the step size of the training algorithm. Thus, as long as the Jacobians have the correct sign, the algorithm will converge if the step size parameter is sufficiently small. For the true prediction error method, the effect of an erroneous forward model is more unclear since past Jacobians enter the calculation of the gradient as well, thus influencing the search direction.

The dead-beat character appearing when inverse models are used directly as controllers will often result in an unnecessarily fast response to reference changes. An active control signal may even harm the system or the actuators. Consequently, it might be desirable to train the network to achieve some prescribed low-pass behavior of the closed-loop system. Say, have the closed-loop system following the model

$$y_m(t) = \frac{B_m(q^{-1})}{A_m(q^{-1})} r(t) \; . \tag{3.22}$$

The polynomials A_m and B_m can be selected arbitrarily by the designer.

This can be obtained by simply using the error $e(t) = y_m(t) - y(t)$ in the criterion (3.15) instead. The control design is in this case related to a Model-Reference Adaptive System (MRAS) (Isermann et al., 1992); a popular type of adaptive controller. Figure 3.7 displays the principle of the specialized training scheme.

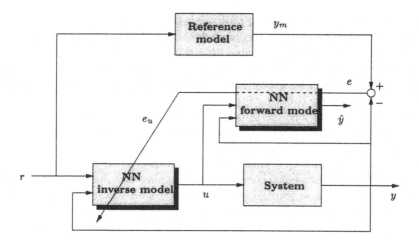

Figure 3.7. Specialized training principle.

Since specialized training is an on-line approach, the combination of having many weights to adjust and having only the slow convergence of a gradient method will often be disastrous. Before the weights are properly adjusted the system may have been driven outside its operating range with possibly serious consequences. Often general training can be used to provide a decent initialization of the network so that specialized training is only used for "fine-tuning" of the controller. This is a highly recommended approach. Since the criteria minimized in the two schemes are different, this does not always solve the problem, however. A rapid initial convergence is often essential and a recursive Gauss-Newton method similar to the one described in Section 2.4 is therefore preferred over a gradient descent algorithm like back-propagation. The algorithm derived for network training in Section 2.4 can be used with only minor revisions.

While proper excitation, optimal model structure selection, and overtraining are important issues in relation to general training, it is mostly of little concern in specialized training. In many applications the system is operated only to perform simple repetitive tasks. For example there is a large number of robotics applications where the same reference trajectory is repeated over and over again. Because specialized training is a direct adaptive control scheme, the network can be trained directly on the relevant reference trajectory. The controller will thus be optimized for this particular trajectory only. Although it is possible to operate the system in other regimes than those specified by the reference, they will not be part of the actual operating range. How the network behaves in these regimes is consequently of little concern. One should simply pursue the smallest possible training error.

Another benefit from performing the training on-line is that the scheme can be applied to systems with time-varying dynamics. Previously it was assumed that the inverse model was generated for a time-invariant system by applying specialized training only until the weights were properly adjusted. However, if the training is never turned off, the scheme should in principle work as an adaptive controller for time-varying systems as well. Naturally the "forward" model providing the Jacobians must be updated on-line, simultaneously with the update of the inverse model. This is a quite subtle application and it is certainly of limited practical relevance. Due to the number of weights typically present in a neural network, this can only be expected to work for systems with very slow variations in the dynamics. A number of numerical robustness issues need to be taken into account when recursive estimation algorithms are applied over long periods of time (Åström and Wittenmark, 1995).

The (simplified) specialized training is quite easily implemented with the back-propagation algorithm derived in Section 2.4. The back-propagation algorithm is used on the inverse model by assuming the following "virtual" error on the output of the controller:

$$e_u(t) = \frac{\partial \hat{y}(t)}{\partial u(t-1)} e(t) . \tag{3.23}$$

If the recursive Gauss-Newton method is implemented, the algorithm in Section 2.4 is used directly except that the partial derivative vector $\psi(t)$ is now defined by

$$\psi(t) = \frac{\partial \hat{y}(t)}{\partial u(t-1)} \phi(t) = \frac{\partial \hat{y}(t)}{\partial u(t-1)} \frac{\partial u(t-1)}{\partial \theta} , \tag{3.24}$$

or for the true recursive prediction error method

$$\psi(t) = \frac{\partial y(t)}{\partial u(t-1)} \psi_u(t) = \frac{\partial y(t)}{\partial u(t-1)} \frac{du(t-1)}{d\theta}$$

$$= \frac{\partial y(t)}{\partial u(t-1)} \left[\phi(t) + \sum_{i=1}^{n} \frac{\partial u(t-1)}{\partial y(t-i)} \psi(t-i) + \right.$$

$$\left. \sum_{i=2}^{m} \frac{\partial u(t-1)}{\partial u(t-i)} \psi_u(t-i) \right] . \tag{3.25}$$

Any of the variations suggested in Section 2.4 can be employed; the exponential forgetting algorithm, the constant trace algorithm, or the EFRA algorithm.

For an MLP-network with one hidden layer of *tanh* units and a linear output

$$\hat{y}(t) = \sum_{j} W_j \tanh \left[\sum_{k} w_{jk} \varphi_k(t) + w_{j0} \right] + W_0 \tag{3.26}$$

the derivative of the output with respect to the regressor $\varphi_i(t)$ is given by

$$\frac{\partial \hat{y}(t)}{\partial \varphi_i(t)} = \sum_j W_j w_{jk} \left(1 - \tanh^2 \left[\sum_k w_{jk} \varphi_k(t) + w_{j0} \right] \right) . \qquad (3.27)$$

An algorithm for specialized training is outlined in Table 3.1.

Table 3.1. Algorithm for specialized training.

1: Read $y(t)$ from A/D-converter.
2: Calculate the error: $e(t) = y_m(t) - y(t)$.
3: Calculate the virtual error:

$$e_u(t) = \frac{\partial \hat{y}(t)}{\partial u(t-1)} e(t)$$

and update weights by back-propagation.
3 alt.: Update weights with the recursive Gauss-Newton method.
4: Evaluate the inverse model to compute the control input $u(t)$. Apply this to the system.
5: Compute the prediction $\hat{y}(t)$ and the Jacobian of the "forward" model. If the recursive Gauss-Newton method is applied determine also $\psi(t+1)$.
6: Go to *Step 1*.

If the computational time delay is significant compared with the sampling period, it is recommended in the specialized training algorithm (see Table 3.1) to relocate the update in *Step 3* to after *Step 4*. This way the entire update takes place between D/A-call and A/D-call.

3.2.4 Specialized Training and Direct Inverse Control of the Benchmark System

To prove the mentioned ability to optimize the controller for a specific reference trajectory, specialized training has been applied to the benchmark system from the previous example. Beginning with the network created using general training, the training process was continued directly on the reference trajectory used in Figure 3.6. The trajectory was repeated eight times while updating the inverse model. Figure 3.8 shows the final controller performance.

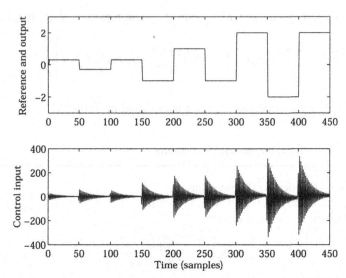

Figure 3.8. Direct inverse control with an inverse model obtained with specialized training. Upper panel: reference and output signal. Lower panel: control signal.

It is evident from the response that perfect model-following was achieved with specialized training. The control signal is still very active, but in Section 3.6 an extension of the specialized training scheme will be introduced that can take care of this.

3.2.5 Section Summary

In this section, the principle of direct inverse control with neural networks was introduced. The scheme was characterized by the controller being a network trained as the inverse of the system. "Inverse" was understood in the sense that the transfer function for the closed-loop system, consisting of controller and system, equalled the time delay of the system. Two different methods for training the network were presented:

Generalized training. In this scheme the network was trained off-line to minimize the mean square error between a control signal applied to the system in an initial experiment and the control signal produced by the neural network. The network could be trained with any of the training methods presented in Chapter 2.

Specialized training. The objective was here to minimize the mean square error between reference signal and the output of the system. This was done (on-line) with a recursive training algorithm.

It was argued that the issue of excitation was of particular concern in relation to generalized training. Due to the fashion in which the training set is collected, it is difficult to avoid that certain regimes of the operating range are not properly represented. Adding to this the problems with generating inverse models of systems not being one-to-one, it was concluded that generalized training should be used mainly for initializing the network. The network is subsequently fine-tuned with specialized training. A complete working procedure for standard problems is outlined in Table 3.2.

Table 3.2. General procedure for training of inverse models.

> **1:** Conduct an experiment to generate a data set.
> **2:** Following the guidelines given in Chapter 2, a "forward" model of the system is identified.
> **3:** Initialize the inverse model with general training. Use for example the Levenberg-Marquardt method.
> **4:** Proceed with specialized training "off-line" by using the model of the system instead of the actual system. Apply a recursive Gauss-Newton algorithm with forgetting for rapid convergence but be careful with "covariance blow-up."
> **5:** Conclude the session by on-line specialized training. Terminate the training algorithm when an acceptable model-following behavior has been achieved.

The major characteristics of direct inverse control are briefly recapitulated below:

Advantages:

- Intuitively simple.

- Simple to implement.

- With specialized training the controller can be optimized for a specific reference trajectory.

- It is (in principle) straightforward to apply specialized training on time-varying systems.

Disadvantages:

- Does not work for systems with an unstable inverse, which unfortunately often occur when using a high sampling frequency.

- Problems for systems not being one-to-one (generalized training of the inverse models).

- Problems with inverse models that are not well-damped (local linearized models will have zeros near the unit circle).

- Lack of tuning options.

- Generally expected to show a high sensitivity to disturbances and noise.

3.3 Internal Model Control (IMC)

Internal model control is a control design closely connected to direct inverse control. It has mainly been used in relation to control of chemical processes, but it can indeed be deployed for other applications too. There are quite restrictive requirements to the characteristics of the system to be controlled, which limits the applicability of IMC a great deal. The limitations are even more restrictive than for direct inverse control. However, IMC has some nice features; e.g., it renders compensation for constant disturbances. The concept of internal model control is treated thoroughly in Morari and Zafiriou (1989), and the idea of using neural networks for IMC was suggested in Hunt and Sbarbaro-Hofer (1991).

3.3.1 Internal Model Control with Neural Networks

An internal model controller requires a forward model as well as a model of the inverse of the system to be controlled. These two models are created by following the instructions given in Chapter 2 and Section 3.2, respectively. The IMC principle is depicted in Figure 3.9. The figure shows an additive disturbance, v, is acting on the output of the system.

In contrast to direct inverse control, the feedback is not composed directly of the system's output. Instead, the error between system output and model output is fed back. Assuming the model is perfect and no disturbances are acting on the system, this feedback signal will be zero. The controller will then be a pure feedforward from the reference. To gain some insight it is useful to calculate different closed-loop transfer functions under linear conditions. Let F, C, P, and M in Figure 3.9 be transfer functions in the delay operator, q^{-1}:

$$y(t) = \frac{q^{-d}FCP}{1 + q^{-d}FC(P - M)}r(t) + \frac{1 - q^{-d}FCM}{1 + q^{-d}FC(P - M)}v(t)$$

$$= v(t) + \frac{q^{-d}FCP}{1 + q^{-d}FC(P - M)}[r(t) - v(t)] \tag{3.28}$$

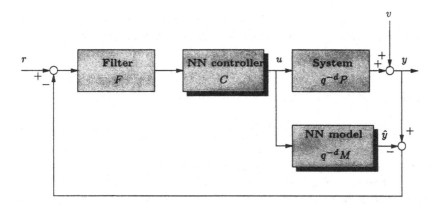

Figure 3.9. Principle of internal model control. The controller is implemented with two neural networks: a model of the system (M) and an inverse model (C). v is a disturbance acting on the output of the system.

and

$$u(t) = \frac{FC}{1 + q^{-d}FC(P - M)} \left[r(t) - v(t)\right] . \tag{3.29}$$

Obviously, a condition for global stability of the closed-loop system is that the system to be controlled and the inverse model are both stable. It is clearly a somewhat restrictive requirement that the system must be open-loop stable, and the class of systems where IMC might come into consideration is thus quite limited.

Under idealized conditions, $M = P$ and $C = P^{-1}$, and the above expressions reduce to

$$u(t) = \frac{F}{P} \left[r(t) - v(t)\right] \tag{3.30}$$

$$y(t) = q^{-d}Fr(t) + \left[1 - q^{-d}F\right] v(t) . \tag{3.31}$$

The low-pass filter, F, obviously has an impact on the behavior of the closed-loop system; on the model-following behavior as well as on the disturbance rejection ability. Hence, the demands to the filter are two-fold:

- It must be stable and have unity stady-state gain to ensure tracking of the reference.

- The numerator of $1 - q^{-d}F$ must contain as a factor the polynomial $\phi(q^{-1})$, with the property $\phi(q^{-1})d(t) = 0$.

By selecting $F = 1$, a dead-beat controller with inherent compensation for constant disturbances is obtained. This can also be seen by determination of the equivalent unity feedback controller. Making a cut at the path leading to the system to be controlled and calculating the transfer function from the error between reference and output and to the control input gives

$$C_e(q^{-1}) = \frac{FC}{1 - q^{-d}FCM} = \frac{FC}{1 - q^{-d}F} \ . \tag{3.32}$$

It is clear that by selecting $F = 1$, integral action is obtained if the time delay is $d = 1$.

The considerations above were made under the assumption of linearity. Regarding the extension to the nonlinear case, the reader is referred to Hunt and Sbarbaro-Hofer (1991) for a treatment on how to use neural networks for IMC. The principle is that the system and its inverse are modelled by a neural network. In Hunt and Sbarbaro-Hofer (1991) it is also explained which assumptions one must make to yield an off-set free response to constant disturbances.

One of the practical problems associated with a neural network implementation of IMC will be to train the networks properly if the disturbance is always present.

3.3.2 Section Summary

This section gave a short introduction to Internal Model Control (IMC). The linear case was presented to illustrate the principle.

IMC requires that the system as well as its inverse are stable, and thus there are many applications where the design cannot be used. The only design parameter is the filter, F. Like for direct inverse control it is therefore difficult to impose constraints on the behavior of the control signal. If the dynamics of the inverse model are not well-damped (corresponding to zeros situated near the unit circle in the linear case), the control signal can be large and oscillating.

Most features of the IMC concept are the same as for direct inverse control. However, the following features are special to the IMC:

- Off-set free response for systems affected by a constant disturbance.
- A requirement that the system is open-loop stable.
- Due to the structure of an IMC-controller it is difficult to ensure that the inverse model is trained on a realistic data set.

3.4 Feedback Linearization

In the area of nonlinear control, feedback linearization is a principle which has drawn much attention. The application is restricted to certain classes of systems, but these are actually not uncommon in practice. The advantage of feedback linearization is that the design can be used generically, in the sense that the same principle can be used on all systems of the right type. Moreover, extensions have been developed to take into account possible model inaccuracies. Design and associated stability analysis is based on quite well-established theory. See for example Slotine and Li (1991), Isidori (1995), and Khalil (1996).

Compared with the amount of work done in the area, the treatment given in the following may occur somewhat superficially, and it will not demonstrate the full potential of the principle. A large amount of publications exists on how to use neural networks for feedback linearization, e.g., Xu et al. (1991), Tzirkel-Hancock and Fallside (1992), Sanner and Slotine (1992), Jin et al. (1992), Yesildirek and Lewis (1994), and Jagannathan and Lewis (1996). The interested reader is encouraged to study these. Common to all the mentioned publications is that continuous models are considered and that the feedback linearization is used in an adaptive control framework. The approach is related to the MRAS approach described in Åström and Wittenmark (1995) in the way stable weight updating laws are derived. Publications on discrete-time approaches are more rare. One example is Chen and Khalil (1991), which proposes a discrete adaptive control scheme. Although such adaptive controllers may work in principle, the practical relevance is often limited due to noise, actuator and output constraints, etc.

The simple strategy presented in this section fits the overall approach taken in this book in that it is a control design based on a discrete input-output model. The presented feedback linearization controller is closely related to the direct inverse controller described earlier.

This section begins by briefly outlining the principle of input-output linearization for continuous models on state space form. A discrete counterpart based on an input-output model structure is then presented. Subsequently it will be shown how to implement this with neural networks. Finally the performance is demonstrated on the benchmark system.

3.4.1 The Basic Principle of Feedback Linearization

Feedback linearization is commonly discussed in a continuous-time framework and this will be the starting point here as well. The fundamental assumption made about the system is that the model of it be written in the canonical form

$$\dot{x}(t) = \begin{bmatrix} \dot{x}_1(t) \\ \dot{x}_2(t) \\ \vdots \\ \dot{x}_n(t) \end{bmatrix} = \begin{bmatrix} x_2(t) \\ x_3(t) \\ \vdots \\ \tilde{f}[x(t)] + \tilde{g}[x(t)]u(t) \end{bmatrix}, \tag{3.33}$$

with $\tilde{f}$ and $\tilde{g}$ being nonlinear functions of the states. t is now the actual time and not a multiple of the sampling period. If this form is not obtained directly when modelling the system, it must be possible to derive it through an appropriate diffeomorphic transformation (Slotine and Li, 1991).

The system can be linearized by introduction of the following control redefinition (it is assumed that $\tilde{g}[x(t)] \neq 0$)

$$u(t) = \frac{w(t) - \tilde{f}[x(t)]}{\tilde{g}[x(t)]} . \tag{3.34}$$

If complete knowledge about the states is available; either from measurements or from an observer, a pole placement type design is easily accomplished. Selecting the *virtual control input*, w, as the reference plus a linear combination of the states results in a closed-loop system specified by

$$\dot{x}(t) = \begin{bmatrix} \dot{x}_1(t) \\ \dot{x}_2(t) \\ \vdots \\ \dot{x}_n(t) \end{bmatrix} = \begin{bmatrix} 0 & 1 & 0 & \cdots & 0 \\ 0 & 0 & 1 & \cdots & 0 \\ \vdots & & & \ddots & \vdots \\ -a_0 & -a_1 & \cdots & & -a_{n-1} \end{bmatrix} x(t) + \begin{bmatrix} 0 \\ 0 \\ \vdots \\ 0 \\ 1 \end{bmatrix} r(t) \tag{3.35}$$

$$y(t) = \begin{bmatrix} 1 & 0 & \cdots & 0 \end{bmatrix} x(t) ,$$

corresponding to the transfer function model

$$H_{cl}(s) = \frac{1}{s^n + a_{n-1}s^{n-1} + \ldots a_0} . \tag{3.36}$$

The coefficients $\{a_i\}_{i=0}^{n-1}$ will specify the characteristic polynomial; i.e., the poles of the closed-loop system. The connection to pole placement control is thus obvious.

Discretization of nonlinear systems is a quite involved action unless it is done by crude approximations. Here a somewhat pragmatic approach will be taken, similar to the one suggested in Chen and Khalil (1991). It is assumed that the system can be modelled as

$$y(t) = f[y(t-1), \ldots, y(t-n), u(t-2), \ldots, u(t-m)]$$
$$+ g[y(t-1), \ldots, y(t-n), u(t-2), \ldots, u(t-m)]u(t-1) \tag{3.37}$$

or equivalently

$$x(t+1) = \begin{bmatrix} x_1(t+1) \\ x_2(t+1) \\ \vdots \\ x_n(t+1) \end{bmatrix} = \begin{bmatrix} x_2(t) \\ x_3(t) \\ \vdots \\ f[\cdot] + g[\cdot]u(t) \end{bmatrix}$$

$$y(t) = x_n(t) , \tag{3.38}$$

with the state vector being defined by

$$x(t) = [y(t - n + 1),\ \dots\ ,y(t - 1),\ y(t)]^T\ . \tag{3.39}$$

Assuming the functions f and g are known, introduction of the following control redefinition will linearize the system at the sampling instants:

$$u(t) = \frac{w(t) - f[y(t),\ \dots\ ,y(t-n+1),u(t-1),\ \dots\ ,u(t-m+1)]}{g[y(t),\ \dots\ ,y(t-n+1),u(t-1),\ \dots\ ,u(t-m+1)]}\ . \tag{3.40}$$

Selecting the virtual control input, w, as the reference plus an appropriate linear combination of past outputs again allows for an arbitrary assignment of the closed-loop poles. The control design can be regarded as a nonlinear counterpart to pole placement with full zero cancellation (see also Section 3.7).

3.4.2 Feedback Linearization Using Neural Network Models

In case the system is unknown, a model can be induced from data by letting two separate neural networks approximate the functions f and g

$$\begin{aligned}
\hat{y}(t|\theta) &= \hat{f}[y(t-1),\ \dots\ ,y(t-n),u(t-2),\ \dots\ ,u(t-m),\theta_f] \\
&+ \hat{g}[y(t-1),\ \dots\ ,y(t-n),u(t-2),\ \dots\ ,u(t-m),\theta_g]u(t-1)\ .
\end{aligned} \tag{3.41}$$

The closed-loop system consisting of controller and system to be controlled is depicted in Figure 3.10.

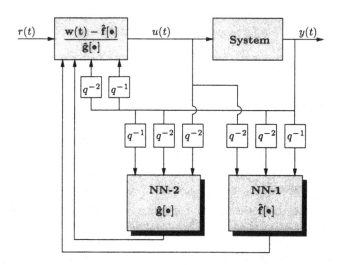

Figure 3.10. Discrete input-output linearization with neural networks.

Derivation of a training method for determination of the weights in the two networks used for approximating f and g is straightforward. The prediction error approach requires knowledge of the derivative of the model output with respect to the weights. In order to calculate this derivative, the derivative of each network output with respect to the weights in the respective network must be determined first

$$\psi_f(t,\theta_f) = \frac{\partial \hat{f}}{\partial \theta_f} \quad \psi_g(t,\theta_g) = \frac{\partial \hat{g}}{\partial \theta_g} \ . \tag{3.42}$$

These derivatives are calculated in the same way as for NNARX models. The derivative of the model output with respect to the weights is then composed of the derivatives of each network in the following manner

$$\psi(t,\theta) = \frac{\partial \hat{y}(t|\theta)}{\partial \theta} = \begin{bmatrix} \psi_f(t,\theta_f) \\ \psi_g(t,\theta_g)u(t-1) \end{bmatrix} \tag{3.43}$$

With this derivative in hand, any of the training methods discussed in Chapter 2 can be used without further modification.

3.4.3 Feedback Linearization of the Benchmark System

To demonstrate the performance of the proposed controller, it has been tested on the benchmark system also used in the previous examples. The model of f is a neural network with 5 *tanh* hidden units, and the model of g is a network with 5 *tanh* hidden units. The controller is designed so that both poles are situated at $z = 0.7$.

Figure 3.11 displays the result of the simulation. For comparison it was performed with the same reference trajectory as considered previously. The simulation reveals the relationship with direct inverse control: close model-following and a very active control signal.

3.4.4 Section Summary

Feedback linearization was proposed as a method for designing pole placement type controllers for a particular class of nonlinear systems. The neural network used for modelling the system must have a specific structure in order to implement the controller.

The controller is related to the model-reference controller discussed previously: a nonlinear controller is designed to make the closed-loop system behave linearly according to a specified transfer function model. This has the consequence that feedback linearization is basically subject to the same limitations as direct inverse control and shares the same weaknesses. Major characteristics of the design are listed below.

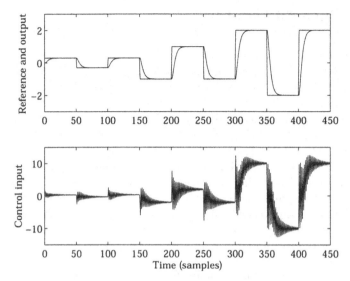

Figure 3.11. Using feedback linearization to control the benchmark system. The closed-loop poles are situated at $z = -0.7$. Upper panel shows reference signal, desired closed-loop response, and actual closed-loop response. Desired and actual response are indistinguishable. Lower panel: the control signal.

Advantages:

- Implementation is simple.

- Only a model of the system to be controlled is required.

- Tuning of the closed-loop response can be made without retraining of the model. The same is, however, possible for the direct inverse controller. An outer feedback can be introduced containing a linear pole placement controller.

Disadvantages:

- Restricted to a particular class of systems. Difficult to resolve whether an unknown system actually belongs to this class.

- Model structure selection complicated because *two* neural network architectures must be chosen.

- Lack of design parameters for tuning of the controller.

- Problems when the inverse (locally) is unstable or near the stability border.

3.5 Feedforward Control

The controllers addressed in the previous sections were all based on the principle of feedback. The main reasons for using feedback are to stabilize unstable systems and to reduce the influence from possible disturbances and model inaccuracies. Using feedback to ensure that the system rapidly follows changes in the reference is not always good practice. A rapid reference tracking obtained with feedback generally has the side effect that the controller becomes highly sensitive to noise. As the previously examined inverse model techniques used feedback to yield a good reference tracking, these designs will often result in a noise sensitive controller with poor robustness properties. To achieve a satisfying reference tracking without feedback, it is instead advised to use a *feedforward* governed only by the reference. The feedforward control should be used for improving the reference tracking while feedback is used for stabilizing the system and for suppressing disturbances.

$$\text{control input} = \text{feedforward} + \text{feedback} \qquad (3.44)$$

Robotics is a good example of an application where rapid reference tracking often is of vital importance; thus, feedforward frequently appears in controllers for robots. The so-called *computed torque* is an example of such a feedforward implementation (Craig, 1989).

Two types of feedforward strategies are proposed in this section: a *dynamic* and a *static*. In a small simulation study it is then demonstrated how feedforward can be used for optimizing the control produced by a conventional PID controller.

3.5.1 Feedforward for Optimizing an Existing Control System

Most industrial systems are today being controlled with PID controllers (including P, PI, and PD). This is done regardless of the system to be controlled being nonlinear and despite the fact that the simplicity of the concept often limits the performance. The reason why PID controllers have gained such popularity is that detailed knowledge about the system is not required, but the controller can be tuned by means of simple rules of thumb. In fact, quite successful PID auto-tuners exist that adjust the design parameters by the push of a button (Åström and Wittenmark, 1995). The control engineer with years of experience with PID controllers thus has much confidence in the design and will be reluctant to substitute it for a more complex control systems like one based on neural networks. As a compromise one might consider enhancing an existing feedback control system, based on a PID controller with a neural network feedforward.

There are different ways in which the feedforward controller can be implemented. The most obvious approach is probably to use a dynamic feedforward. That is, an inverse model is trained by generalized or specialized training as discussed in Section 3.2:

$$u(t) = \hat{g}[y(t+d), \ \dots \ , y(t+d-n), u(t-1), \ \dots \ , u(t-m+1)] \ . \qquad (3.45)$$

The feedforward component of the control input is then composed by substituting all system outputs for corresponding reference values (Madsen, 1995):

$$u_{ff}(t) = \hat{g}[r(t+d), \ \dots \ , r(t+d-n),$$
$$u_{ff}(t-1), \ \dots \ , u_{ff}(t-m+1)] \ . \qquad (3.46)$$

The principle of additive feedforward is shown in Figure 2.12. If the inverse model is stable, the introduction of a feedforward controller will not change the stability properties of the closed-loop system. However, it might be difficult to resolve whether or not the inverse model is in fact stable.

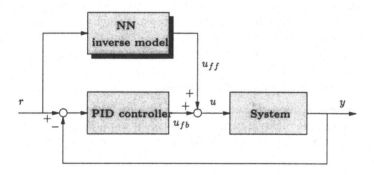

Figure 3.12. Optimization of an existing control system by adding a neural network synthesized feedforward signal based on the reference signal.

In case the complete reference trajectory is known in advance, implementation of the scheme is particularly easy. It is then possible to compute the contribution from the feedforward controller beforehand and store the entire sequence of control inputs $\{u_{ff}\}$ for use in the computer program implementing the control system.

The concept has more attractive features of practical relevance. Since it is assumed that a stabilizing controller is available in advance, the experiment conducted to collect a set of training data set is easily performed. Another advantage is that one can introduce the feedforward signal gradually. In applications where an inappropriate control input can cause damages, this is a comfortable strategy.

A useful alternative to the dynamic feedforward is to use a static feedforward. This concept completely overcomes the potential stability problems with the dynamic models. The idea is to use an NNFIR model for describing the inverse of the system.

$$u_{ff}(t) = \hat{g}[r(t+d), \ \dots \ , r(t+d-n_{ff})] \ . \tag{3.47}$$

Although one cannot model a dynamic system accurately by a FIR-type model, NNFIR-models are attractive because they do not contain feedback. The modelling errors need not have a harmful impact on the control. Due to the presence of a conventional feedback controller, a slightly erroneous feedforward signal will be compensated.

Sometimes feedforward control is also used in regulation problems where the reference attains constant levels for longer periods of time. To speed up the tracking of set-point changes, a feedforward controller is typically designed to provide the steady-state value of the control signal (Franklin et al., 1998). By conducting the initial experiment specifically to reach a number of different steady states over the entire range of outputs, a network can subsequently be trained on the stored steady-state relationships. If a network has learned the inverse correspondence between the steady-state control inputs and outputs, it is straightforward to use it for feedforward purposes. The steady-state feedforward is not suitable for unstable systems as in this case the control input should always be zero in steady state.

Although a neural network feedforward can be useful for optimizing many control systems, one must be careful not to use it uncritically. An inaccurate feedforward control may actually harm rather than enhancing performance.

3.5.2 Feedforward Control of the Benchmark System

To demonstrate the usefulness of the feedforward concept, it is applied to a set-up where the benchmark system is being stabilized by a discretized version of the PID controller:

$$D(s) = K \frac{1+\tau_I s}{\tau_I s} \frac{1+\tau_D s}{1+\alpha \tau_D s} = 8 \frac{1+5s}{5s} \frac{1+0.8s}{1+0.08s} \ . \tag{3.48}$$

With the PID controller alone, the performance of the closed-loop system displayed in Figure 3.13 was obtained. As the controller was designed for the linearized model for the rest state, the performance is best for small step changes.

Adding a (dynamic) feedforward by using the inverse model created with specialized training in the example in Section 3.2.4 gives the result shown in Figure 3.14.

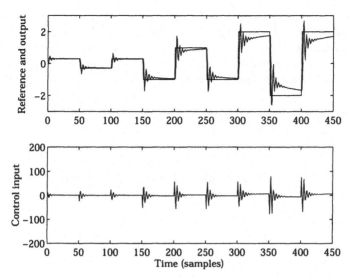

Figure 3.13. Benchmark system controlled by a PID controller. Upper panel: Reference and output. Lower panel: Control signal.

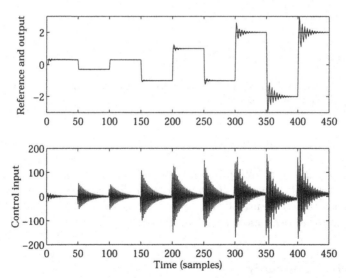

Figure 3.14. Benchmark system controlled using PID control with feedforward from the reference. Upper panel: Reference and output. Lower panel: Control signal.

The reference tracking is obviously improved by introduction of the feedforward. Ideally, the PID controller's contribution to the control signal should be close to zero when the feedforward is introduced. To check this, the contributions from feedforward and PID controller, respectively, are shown in Figure 3.15. It appears that the hypothesis is valid for small steps in the

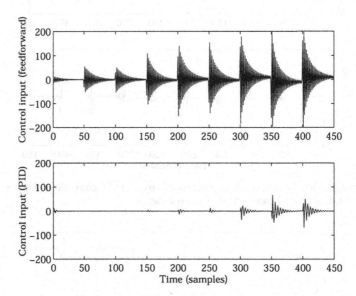

Figure 3.15. The individual contributions to the control signal from feedforward (top) and PID controller (bottom).

reference. As the magnitude of the steps is increased, it becomes visible that the inverse model is not quite accurate. Due to the undamped nature of the inverse model, this application is probably not the most suitable for demonstrating the present type of feedforward control. The suggested concept where a network is trained to predict the steady-state value of the control signal might in fact have been more appropriate here.

3.5.3 Section Summary

In this section it was suggested to add a feedforward control input from an inverse neural network model for enhancing an existing control system. Three different types of feedforward control were described:

Dynamic feedforward. The feedforward controller is an inverse model trained as described in Section 3.2.

Static feedforward. The feedforward controller is an NNFIR type mode, only governed by the reference. As there is no feedback from the output, there will not be problems with stability. Inevitable modelling errors are compensated by the feedback controller.

Steady-state feedforward. The feedforward is a function of the steady-state reference. The function returns the steady-state gain of the inverse model multiplied by the reference. This feedforward should be used in regulation problems with occasional changes in set-point.

The concept of optimizing existing control systems may have more commercial appeal than the other neural-network-based feedback controllers discussed in this book. It concentrates on improving the ability to track the reference rather than changing the closed-loop dynamics. Features like easy implementation, gradual introduction of the feedforward signal, and that the existing control system is kept, make the transition to a neural-network-based control system less dramatic.

Advantages and disadvantages of the neural-network-based feedforward are briefly listed below:

Advantages:

- Easy to implement.
- Can be introduced gradually.
- Improves reference tracking without increasing sensitivity to noise.
- Experiment easily performed if a stabilizing feedback controller is available already.

Disadvantages:

- A requirement that a feedback controller is present.
- Does not reduce the effect of disturbances acting on the system.
- Problems if the inverse is unstable or poorly damped.
- A poor feedforward might harm the performance instead of enhancing it.

3.6 Optimal Control

After having examined a number of controllers all relating to pole placement with full zero cancellation, it is apparent that the strategies had problems for systems with poorly damped or unstable inverses. This was demonstrated in all the examples presented thus far. What is needed is a feature for better "tuning" the shape of the control signal. The idea behind optimal control is to design the controller according to a criterion where reference tracking is rewarded while there is a penalty on the magnitude of (some function of) the control input. The design is implemented by training a neural network to provide the control inputs that minimize this criterion. In this section it will be shown how a simple optimal controller can be trained by straightforward modification of the specialized training algorithm. More advanced concepts can be implemented along the same lines.

3.6.1 Training of an Optimal Controller

The specialized training algorithm discussed in Section 3.2 was attractive because it was *goal directed*. This rendered it useful for training a controller for a specific reference trajectory. This property will now be carried over to an optimal control design. Recall that the specialized training algorithm was derived to minimize the criterion

$$J(\theta) = \sum_t [r(t) - y(t)]^2 \tag{3.49}$$

on-line with a recursive training algorithm. By adding a term for penalizing the squared controls

$$J(\theta) = \sum_t [r(t) - y(t)]^2 + \rho u^2(t) \quad \rho \geq 0 \tag{3.50}$$

a simple type of optimal control criterion is achieved. If $\rho = 0$ the scheme coincides with conventional specialized training, and the controller will become the inverse of the system. As ρ is increased, a "detuned" inverse is obtained instead. The behavior of the controller will deviate increasingly from a pure dead-beat controller while the control signal becomes more smooth and attains smaller values.

As for specialized training, the optimal controller is trained on-line. For example with a recursive back-propagation or Gauss-Newton algorithm. For rapid convergence the latter is typically preferred. Unfortunately the new penalty term has the effect that the criterion is no longer a mean-square error criterion, which is the type of criterion for which the Gauss-Newton algorithm was previously derived. Nevertheless, it is possible to derive a somewhat *ad hoc* modification of the algorithm which complies with the modified criterion and for which practical experience indicates that the convergence properties are quite good. The basic assumption is again (if no forgetting is used):

$$J_t(\theta, Z^t) = J_{t-1}(\theta, Z^{t-1}) + [r(t) - y(t)]^2 + \rho u^2(t-1) \tag{3.51}$$

where J_{t-1} is assumed to be minimized already. Notice that the difference between J_t and J_{t-1} contains $u(t-1)$ and not $u(t)$. This is because the usual time delay of one sampling period has been assumed, in which case $u(t-1)$ is the most recent control input that influences $y(t)$.

Relying on the same approximations as in specialized training, the following expression is used for the derivative of the second term on the right-hand side with respect to the network weights

$$G(\theta) \simeq \frac{du(t-1)}{d\theta}\left[-\frac{\partial\hat{y}(t)}{\partial u(t-1)}e(t) + \rho u(t-1)\right] . \qquad (3.52)$$

For abbreviation introduce again

$$e_u(t) = \frac{\partial\hat{y}(t)}{\partial u(t-1)}e(t) \qquad (3.53)$$

$$\psi_u(t) = \frac{du(t-1)}{d\theta} \qquad (3.54)$$

$$\psi(t) = \frac{\partial\hat{y}(t)}{\partial u(t-1)}\psi_u(t) . \qquad (3.55)$$

With these definitions (3.52) can be written in a more compact form

$$G(\theta) \simeq \psi_u(t)\left[-e_u(t) + \rho u(t-1)\right] . \qquad (3.56)$$

It is not quite clear how the Hessian update should be modified, but experience has shown that this is of little importance. The same type of updates that were used in the specialized training algorithms can be used here without further ado. If, for example, a constant forgetting factor is used, the weights are updated according to

$$P(t) = \left[P(t-1) - \frac{P(t-1)\psi(t)\psi^T(t)P(t-1)}{\lambda + \psi^T(t)P(t-1)\psi(t)}\right]\frac{1}{\lambda} \qquad (3.57)$$

$$\hat{\theta}(t) = \hat{\theta}(t-1) + P(t)\psi_u(t)\left[e_u(t) - \rho u(t-1)\right] . \qquad (3.58)$$

As discussed in the presentation of specialized training, $\psi_u(t)$ can be approximated by the partial derivative in the spirit of pseudo-linear regressions

$$\psi_u(t) \simeq \frac{\partial u(t-1)}{\partial\theta} . \qquad (3.59)$$

The criterion under consideration is an *infinite horizon* type criterion. As discussed previously, it is often interesting to optimize the controller for a specific (finite) reference trajectory. This can be done by repeating simply the trajectory until the weights have converged. This is in some sense related to a minimization of

$$J(\theta) = \mathbf{E}\left\{\sum_{t=1}^{N}[r(t) - y(t)]^2 + \rho u^2(t)\right\}, \qquad (3.60)$$

where N specifies the length of the reference trajectory. An advantage of training the network in this fashion is that the convergence of the algorithm is more easily supervised. One can check that the sum actually decreases each time the reference trajectory is applied.

One particular property of criterion-based controller design for nonlinear systems deserves some attention. While the controllers described in the previous sections all were designed to make the closed-loop system behave *linearly* at the sampling instants, this is not the case for the optimal controller. The closed-loop system may in fact behave quite differently in different regimes of the operating range. A comprehensive test of the closed-loop system is therefore recommended to ensure that the controller behaves satisfactorily over the entire operating range.

Being an extension to specialized training, the implementation should follow the guidelines outlined in Section 3.2. A proper initialization of the controller is perhaps the main problem with the scheme. As suggested for the specialized training algorithm, it may be a good idea to initialize the network off-line by simulation. This is done by simulating the proposed algorithm with the model of the system as a substitute for the real system. A proper value of the penalty factor, ρ, can be determined at this stage as well.

3.6.2 Optimal Control of the Benchmark System

The experiment with specialized training reported in Section 3.2 has been repeated with the proposed optimal controller training algorithm. The factor penalizing the squared controls was set to $\rho = 6 \times 10^{-4}$, and the network was trained with a simplified Gauss-Newton algorithm. The reference trajectory of 450 samples was repeated eight times while updating the weights. A simulation with the trained controller network is displayed in Figure 3.16.

It is seen that the reference tracking ability is deteriorated when compared with direct inverse control: there is some overshoot, and the response has a steady-state error. However, the control signal is smoother and does not attain as large values as in direct inverse control. Overshoot and steady-state error could be made smaller at the expense of a more active control signal by reducing the penalty factor. To avoid the steady-state error for an arbitrary penalty factor, one would have to penalize $[u(t) - u(t-1)]^2$ instead of the squared control input. This will also give a smoother control signal. The criterion-based designs presented later in this chapter will all contain a penalty on the differenced control input.

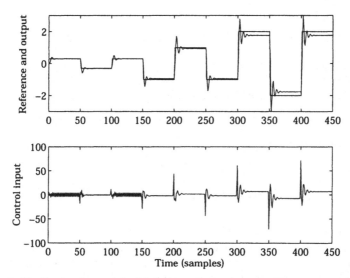

Figure 3.16. Optimal control of benchmark the system. Upper panel: reference and output signals. Lower panel: control signal.

3.6.3 Section Summary

A simple optimal control scheme was presented in this section. The controller was implemented as an extension to specialized training of inverse model controllers. The extension consisted of a second term in the criterion for penalizing squared control inputs. This made the control signal smoother and reduced the magnitude of the signal. The price paid for this was a deterioration in the reference tracking.

A major drawback of the method was that it was necessary to retrain the network each time the penalty factor had been modified. This is an inherent problem with direct design as the controller is implemented by a neural network. The two following sections, Section 3.7 and Section 3.8, will present *indirect designs* in which the controller is not a neural network. This will provide a much higher flexibility in the choice of criteria for criterion-based control design.

The main features of optimal control are listed below:

Advantages:

- Easy to tune.
- Performs well for a large class of systems (compared with direct inverse control).
- Goal directed.

- Suitable for designing a controller for a specific reference trajectory.
- It is (in principle) directly applicable to time-varying systems.

Disadvantages:

- Controller network is trained on-line.
- The network must be retrained whenever the penalty factor is modified.
- Initialization of the network is difficult.

3.7 Controllers Based on Instantaneous Linearization

Linearization of nonlinear models is a technique often used to alleviate the design of controllers for nonlinear systems. When modelling a system deductively (i.e., by white-box modelling), the type of description one arrives at will typically be in terms of a set of nonlinear differential equations. Unless one pursues directly a true nonlinear design method, this differential equation model is linearized around one or more stationary points followed by application of a linear controller design. Although a model obtained by linearization at a first glance may appear to be a crude description of the true system, the advantage of being able to apply the rich collection of well-understood linear design techniques will often compensate for this in terms of performance of the obtained closed-loop system.

It is evident that a model obtained through linearization around an operating point can be considered valid only in a certain regime around this point. The character of the nonlinearities and the size of the operating range will then determine whether it is sufficient to use a single linear model, or it is necessary to obtain more by linearization around a larger set of operating points. In the latter case, the control system must accordingly consist of a bank of controllers as well as some rules for how to switch between them. A control system where one chooses from a set of predefined linear controllers, each having been tuned for a specific operating regime, is commonly referred to as *gain scheduling*. Gain scheduling can be implemented in various ways according to the nature of the application under consideration (Åström and Wittenmark, 1995). In some cases it includes a switching between a finite set of controllers each designed for a particular regime while in other cases the switching progresses continuously. The former implementation is said to have a *finite schedule* while the latter has an *infinite schedule*. The control system design obviously becomes quite complex when it is necessary to design controllers for a number of regimes as well as implementing a detection of the current operating conditions. Typically gain scheduling is therefore considered only under circumstances where a satisfactory control performance cannot be achieved by a single linear controller.

If a nonlinear system is unknown, and a model has to be inferred from a set of experimental data, a conventional linear identification technique will often fail. Sometimes one can clearly distinguish between different operating regimes inside which the system behaves linearly. In these rare cases, the experiment can be conducted so as to excite the individual regimes separately; however, this will require considerable knowledge about the dynamics of the system. In most cases a nonlinear black-box approach, possibly including neural networks, is preferred as this is the most simple.

By using the neural network model structures considered in Chapter 2, discrete nonlinear models are obtained directly. This case is rarely treated in standard textbooks on digital control system design. The reason for this being that most literature on control assumes models obtained by white-box modelling. Thus, a nonlinear model is typically described in terms of differential equations. Usually the model is in this case linearized *before* the discretization and not the other way around. Linearization of discrete nonlinear models can, however, be carried out along the same lines. In this section a linearization technique called *instantaneous linearization* will be applied to the neural network models. The linearization is carried out at each sampling instant and will be used in a gain scheduling type controller. The control designs suggested are based on pole placement and minimum variance designs. To demonstrate the power of the strategy, two different pole placement controllers are made for the benchmark system.

Ideas similar to those presented in this section have been presented in for example Sørensen (1993), Sørensen (1996), Ahmed and Tasadduq (1994), and Lightbody and Irwin (1995).

3.7.1 Instantaneous Linearization

The idea behind instantaneous linearization is to extract a linear model from the nonlinear neural network model at each sample. The principle will be presented for deterministic models first and subsequently it will be extended to cover stochastic models.

Instantaneous Linearization of a Deterministic Model. Assume that a deterministic neural network input-output model of the system to be controlled is available

$$y(t) = g[\varphi(t)] \qquad (3.61)$$

where the regression vector is given by

$$\varphi(t) = [y(t-1), \ \dots \ , y(t-n), u(t-d), \ \dots \ , u(t-d-m)]^T \ . \qquad (3.62)$$

The principle of instantaneous linearization is as follows:

Interpret the regression vector as a vector defining the state of the system and at time $t = \tau$ linearize g around the current state $\varphi(\tau)$ to obtain the approximate model

$$\tilde{y}(t) = -a_1\tilde{y}(t-1) - \ldots - a_n\tilde{y}(t-n)$$
$$+ b_0\tilde{u}(t-d) + \ldots + \tilde{u}(t-d-m), \qquad (3.63)$$

where

$$a_i = -\left.\frac{\partial g[\varphi(t)]}{\partial y(t-i)}\right|_{\varphi(t)=\varphi(\tau)} \qquad b_i = \left.\frac{\partial g[\varphi(t)]}{\partial u(t-d-i)}\right|_{\varphi(t)=\varphi(\tau)} \qquad (3.64)$$

and

$$\tilde{y}(t-i) = y(t-i) - y(\tau-i) \qquad \tilde{u}(t-i) = u(t-i) - u(\tau-i). \qquad (3.65)$$

Separating from the rest the part of the expression containing components of the current regression vector, $\varphi(\tau)$, the approximate model can alternatively be expressed as

$$y(t) = \left[1 - A(q^{-1})\right] y(t) + q^{-d}B(q^{-1})u(t) + \zeta(\tau), \qquad (3.66)$$

where the bias term, $\zeta(\tau)$, is given by

$$\zeta(\tau) = y(\tau) + a_1 y(\tau-1) + \ldots + a_n y(\tau-n)$$
$$-b_0 u(\tau-d) - \ldots - b_m u(\tau-d-m). \qquad (3.67)$$

The coefficients $\{a_i\}$ and $\{b_i\}$ have been collected in the polynomials $A(q^{-1})$ and $B(q^{-1})$,

$$A(q^{-1}) = 1 + a_1 q^{-1} + \ldots + a_n q^{-n} \qquad (3.68)$$
$$B(q^{-1}) = b_0 + b_1 q^{-1} + \ldots + b_m q^{-m}. \qquad (3.69)$$

The approximate model may thus be interpreted as a linear model affected by a constant disturbance, $\zeta(\tau)$, depending on the current operating point.

For an MLP-network with one hidden layer of *tanh* units and a linear output,

$$\hat{y}(t) = \sum_{j=1}^{n_h} W_j \tanh\left[\sum_{k=1}^{n_\phi} w_{jk}\varphi_k(t) + w_{j0}\right] + W_0, \qquad (3.70)$$

the derivative of the output with respect to input $\varphi_i(t)$ is calculated in accordance with

$$\frac{\partial \hat{y}(t)}{\partial \varphi_i(t)} = \sum_{j=1}^{n_h} W_j w_{jk}\left(1 - \tanh^2\left[\sum_{k=1}^{n_\phi} w_{jk}\varphi_k(t) + w_{j0}\right]\right). \qquad (3.71)$$

Instantaneous Linearization of a Stochastic Model. It is straightforward to apply instantaneous linearization to the model structures used for describing systems affected by stochastic disturbances (Sørensen, 1996). It makes little difference whether the models are specified in input-output form or if state space descriptions are being used.

Consider for example the NNARMAX-model as a representative for the family of input-output models. The network renders in this case a one-step ahead predictor for the system in the following fashion

$$\hat{y}(t) = g[\varphi(t)] . \tag{3.72}$$

The regression (or state) vector is now

$$\varphi(t) = [y(t-1), \; ... \; , y(t-n), \; u(t-d), \; ... \; , u(t-d-m),$$
$$\varepsilon(t-1), \; ... \; , \varepsilon(t-k)]^T . \tag{3.73}$$

Similarly, introduce the following notation for the partial derivatives of the predictor with respect to each regressor

$$a_i = -\left.\frac{\partial g[\varphi(t)]}{\partial y(t-i)}\right|_{\varphi(t)=\varphi(\tau)} \qquad b_i = \left.\frac{\partial g[\varphi(t)]}{\partial u(t-d-i)}\right|_{\varphi(t)=\varphi(\tau)} \tag{3.74}$$

$$c_i = \left.\frac{\partial g[\varphi(t)]}{\partial \varepsilon(t-i)}\right|_{\varphi(t)=\varphi(\tau)} .$$

The linearized prediction model then takes the form

$$\hat{y}(t) = \left[1 - A(q^{-1})\right] y(t) + q^{-d}B(q^{-1})u(t)$$
$$+ \left[C(q^{-1}) - 1\right]\varepsilon(t) + \zeta(\tau) , \tag{3.75}$$

with the bias term defined by

$$\zeta(\tau) = y(\tau) + a_1 y(\tau-1) + \; ... \; + a_n y(\tau-n)$$
$$-b_0 u(\tau-d) - \; ... \; - b_m u(\tau-d-m)$$
$$+c_1 \varepsilon(\tau-1) + \; ... \; + c_k \varepsilon(\tau-k) \tag{3.76}$$

and

$$C(q^{-1}) = 1 + c_1 q^{-1} + \; ... \; + c_k q^{-k} . \tag{3.77}$$

From the discussion of optimal predictors for linear systems in Section 2.2, recall that the predictor (3.77) corresponds to the optimal predictor for an ARMAX model

$$A(q^{-1})y(t) = q^k B(q^{-1})u(t) + C(q^{-1})e(t) + \zeta(\tau) . \tag{3.78}$$

$\{e(t)\}$ is a white noise sequence independent of past control inputs.

Nonlinear models on state space innovation form can be linearized in a completely analogous fashion (x, y, u, ε are here vectors):

$$\hat{x}(t) = g[\varphi(t)] \tag{3.79}$$

$$\hat{y}(t) = C\hat{x}(t) \tag{3.80}$$

where

$$\varphi(t) = \begin{bmatrix} \hat{x}(t-1) \\ u(t-1) \\ \varepsilon(t-1) \end{bmatrix} \tag{3.81}$$

and defining the matrices A, B and K

$$A = \left.\frac{\partial g[\varphi(t)]}{\partial x(t)}\right|_{\varphi(t)=\varphi(\tau)} \qquad B = \left.\frac{\partial g[\varphi(t)]}{\partial u(t)}\right|_{\varphi(t)=\varphi(\tau)} \tag{3.82}$$

$$K = \left.\frac{\partial g([\varphi(t)]}{\partial \varepsilon(t)}\right|_{\varphi(t)=\varphi(\tau)},$$

the linearized predictor is then given by

$$\hat{x}(t) = A\hat{x}(t-1) + Bu(t-1) + K\varepsilon(t-1) + \Phi(\tau) \tag{3.83}$$

$$\hat{y}(t) = C\hat{x}(t) \tag{3.84}$$

where

$$\Phi(\tau) = \hat{x}(\tau) - A\hat{x}(\tau-1) - Bu(\tau-1) - K\varepsilon(\tau-1). \tag{3.85}$$

This is also known as the *Kalman filter*, which is the optimal predictor for the linear system

$$x(t) = Ax(t-1) + Bu(t-1) + Ke(t-1) + \Phi(\tau) \tag{3.86}$$

$$y(t) = Cx(t) + e(t). \tag{3.87}$$

3.7.2 Applying Instantaneous Linearization to Control

Application of the instantaneous linearization technique to design of controllers is obvious: at each sample a linear model is extracted from a neural network model of the system and a linear controller is designed. The concept is illustrated in Figure 3.17. One can regard this as a gain scheduling controller with an infinite schedule.

The structural equivalence with the *indirect self-tuning regulator* examined in several textbooks on adaptive control (such as Åström and Wittenmark (1995)) is evident. In the self-tuning regulator, a recursive estimation algorithm is used for identification of a new linear model at each sampling instant. In this case, however, a linear model is continuously extracted from a more

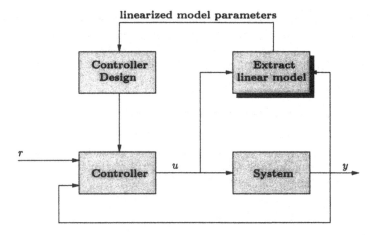

Figure 3.17. Control system based on instantaneous linearization of a neural network model.

complex, nonlinear neural network model. Apart from that, the concepts are similar. As in the self-tuning regulator, the control design is based on the *certainty equivalence principle*; the design block assumes the extracted linear model is a perfect description of the system. Since the concepts show many similarities, many of the considerations can be directly carried over from the self-tuning regulator when selecting the controller design. Likewise, the required computational effort for the two methods is also comparable. Unless the neural network model is extremely large, the most time-consuming task is typically the controller design since this most often includes solving a Diophantine or a Ricatti equation.

A truly appealing feature of the linearization is that essentially any linear control design can be incorporated in the design block. This has a number of advantages, such as stabilization of a large class of systems, easy tuning, and compensation for disturbances. However, one must keep in mind that the design is also subject to limitations. The linearization results in the bias term, $\zeta(\tau)$, which has to be compensated for. This can be achieved by enforcing integral action into the controller in which case other constant disturbances will be compensated for as well. Another limitation is that one shall be careful not to violate the certainty equivalence assumption. Depending on the character of the nonlinearities of the system and the selected reference trajectory, the linear model in general can be considered valid only within a relatively short time horizon. Thus, one must be careful not to rely too heavily on the linearized model and only choose controller designs not violating the limitations of the approximation. A violation might occur by demanding a very fast response to reference changes, in which case the operating region shifts abruptly.

In the following, a few examples of linear controller designs are given, which may be used in combination with instantaneous linearization. First, different variations on *pole placement design* are presented. Subsequently, the so-called *minimum variance controllers* are described.

3.7.3 Approximate Pole Placement Design

It is straightforward to apply the instantaneous linearization technique to obtain an approximate pole placement design. Let the starting point be a linearized deterministic model (or NNARX model)

$$y(t) = \left[1 - A(q^{-1})\right] y(t) + q^{-d}B(q^{-1})u(t) + \zeta(\tau) . \qquad (3.88)$$

Without loss of generality it is assumed that $\deg(A) = d + \deg(B)$. If necessary, $A(q^{-1})$ is extended with the necessary number of zero terms to comply with the condition.

The objective of pole placement is to select the three polynomials R, S, and T in the controller structure shown in Figure 3.18 so that the closed-loop system will behave as a prescribed transfer function model

$$y_m(t) = q^{-d} \frac{B_m(q^{-1})}{A_m(q^{-1})} r(t) . \qquad (3.89)$$

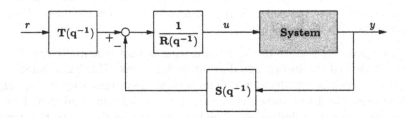

Figure 3.18. The RST controller: $R(q^{-1})u(t) = T(q^{-1})r(t) - S(q^{-1})y(t)$.

The denominator, A_m, which is assumed to be a monic polynomial, is a design parameter for specifying the poles of the desired closed-loop system, thus motivating the names *pole placement* and *pole assignment*. Many different variations on pole placement design exist. If, for example, the polynomials B_m and A_m are to be chosen freely, the design includes a cancellation of the system's zeros. Hence, this variation on pole placement is referred to as *the cancellation controller*. The opposite strategy is to keep all the zeros and only place the poles. A compromise where only some of the zeros are canceled is also possible, but this is more complicated to implement as a factorization of the numerator must be incorporated (Åström and Wittenmark, 1995). Below a review of basic pole placement design will be given with full zero cancellation and with no zeros cancelled. Furthermore, it is discussed how to design a controller for compensation of disturbances with known characteristics.

Pole placement without zero cancellation. Pole placement is perhaps most easily understood by using, as a starting point, estimation-based state space control (see for example Franklin et al. (1998)). In a state space design, the state feedback is designed to situate the poles at prescribed locations. In the case that not all states are measured, an observer is introduced to estimate (at least) the missing states. In the input-output approach taken here, only one measurement is available: the output, $y(t)$. Obviously it is therefore necessary to introduce some sort of observer when the order of the system model exceeds one. The characteristic polynomial of the closed-loop system, A_c, is thus composed of the desired closed-loop model polynomial, A_m, together with the observer polynomial, A_o:

$$H_c(q^{-1}) = q^{-d}\frac{B_c(q^{-1})}{A_c(q^{-1})} = q^{-d}\frac{B_m(q^{-1})A_o(q^{-1})}{A_m(q^{-1})A_o(q^{-1})} . \tag{3.90}$$

Only the *minimum degree solution* will be considered here. That is, no additional poles are introduced in A_m except for an integrator to compensate for $\zeta(\tau)$

$$\deg(A_m) = \deg(A) + 1 \qquad\qquad \deg(A_o) = \deg(A) - 1 . \tag{3.91}$$

The second constraint corresponds to only the non-measured states being estimated. Selecting the minimum degree solution is partially motivated by the desire to perform the design on-line: the amount of computations increases dramatically with the introduction of additional poles. Moreover, the choice of controller polynomials (R, S and T) will not be unique if additional poles are introduced: different solutions may lead to the same closed-loop model.

Applying instantaneous linearization to the nonlinear neural network model and enforcing integral action into the controller implies that the following identity must be solved with respect to R, S and T

$$\frac{B_m(q^{-1})A_o(q^{-1})}{A_m(q^{-1})A_o(q^{-1})} = \frac{B(q^{-1})T(q^{-1})}{\Delta A(q^{-1})\bar{R}(q^{-1}) + q^{-d}B(q^{-1})S(q^{-1})} , \tag{3.92}$$

where

$$R(q^{-1}) = \Delta\bar{R}(q^{-1}) . \tag{3.93}$$

Δ is the difference operator, $\Delta = 1 - q^{-1}$.

In pole placement without zero cancellation the choice of B_m reflects that the zeros are left unchanged

$$B_m(q^{-1}) = \frac{A_m(1)}{B(1)} B(q^{-1}) . \tag{3.94}$$

The polynomial, $T(q^{-1})$ is thus determined by

$$T(q^{-1}) = \frac{A_m(1)}{B(1)} A_o(q^{-1}) , \tag{3.95}$$

and should guarantee a unit steady-state gain.

$\bar{R}(q^{-1})$ and $S(q^{-1})$ are determined from

$$A_m(q^{-1})A_o(q^{-1}) = \Delta A(q^{-1})\bar{R}(q^{-1}) + q^{-d}B(q^{-1})S(q^{-1}) , \tag{3.96}$$

where (assuming $\deg(A) = \deg(B) + d$)

$$\deg(\bar{R}) = \deg(S) - 1 = \deg(A) - 1 . \tag{3.97}$$

The design parameters in pole placement without zero cancellation are the two polynomials A_m and A_o, which should be monic and have their roots well inside the unit circle. Since the zeros are not canceled they will not enter the closed-loop characteristic polynomial. It is therefore possible to control systems with an unstable inverse (zeros outside the unit circle) by using this design method.

The identity (3.96) is called a *Diophantine equation*. By multiplying the polynomials on both sides of the equality sign and comparing the coefficients, a system of linear equations is obtained. For the general Diophantine equation

$$\bar{C}(q^{-1}) = \bar{A}(q^{-1})\bar{R}(q^{-1}) + q^{-d}\bar{B}(q^{-1})\bar{S}(q^{-1}) , \tag{3.98}$$

where

$$\begin{aligned}
\bar{A}(q^{-1}) &= 1 + a_1 q^{-1} + \dots + a_{n_a} q^{-n_a} \\
q^{-d}\bar{B}(q^{-1}) &= b_0 + b_1 q^{-1} + \dots + b_{n_b} q^{-n_b} \\
\bar{C}(q^{-1}) &= 1 + c_1 q^{-1} + \dots + c_{n_c} q^{-n_c} \\
\bar{R}(q^{-1}) &= r_0 + r_1 q^{-1} + \dots + r_{n_r} q^{-n_r} \\
\bar{S}(q^{-1}) &= s_0 + s_1 q^{-1} + \dots + s_{n_s} q^{-n_s} .
\end{aligned}$$

The solution of the Diophantine equation is not unique. If $\{\bar{R}_0, \bar{S}_0\}$ is a solution then

$$\bar{R}(q^{-1}) = \bar{R}_0(q^{-1}) + \bar{B}(q^{-1})F(q^{-1}) \tag{3.99}$$
$$\bar{S}(q^{-1}) = \bar{S}_0(q^{-1}) - \bar{A}(q^{-1})F(q^{-1}) \tag{3.100}$$

is also a solution for an arbitrary polynomial F. Generally, the lowest-order solution is used; the so-called *minimum degree solution*:

$$n_r = n_b - 1 , \qquad n_s = \max(n_a - 1, n_c - n_b) . \qquad (3.101)$$

The system of equations that must be solved to determine $\bar{R}$ and $\bar{S}$ takes the form

$$
\underbrace{\begin{bmatrix}
1 & 0 & \cdots & 0 \\
a_1 & 1 & \ddots & \vdots \\
a_2 & a_1 & & 0 \\
\vdots & \vdots & & 1 \\
a_n & a_{n-1} & & a_1 \\
0 & a_n & & \vdots \\
\vdots & \vdots & & a_{n-1} \\
0 & 0 & & a_n
\end{bmatrix}}_{n_r+1}
\underbrace{\begin{bmatrix}
b_0 & 0 & \cdots & 0 \\
b_1 & b_0 & & \vdots \\
b_2 & b_1 & & b_0 \\
\vdots & b_2 & & b_1 \\
b_{n_b} & \vdots & & b_2 \\
0 & b_{n_b-1} & & \vdots \\
\vdots & b_{n_b} & & b_{n_b-1} \\
0 & \vdots & & b_{n_b}
\end{bmatrix}}_{n_s+1}
\begin{bmatrix}
r_0 \\ r_1 \\ \vdots \\ r_{n_r} \\ s_0 \\ s_1 \\ \vdots \\ s_{n_s}
\end{bmatrix}
=
\begin{bmatrix}
1 \\ c_1 \\ \vdots \\ c_{n_c} \\ 0 \\ \vdots \\ 0
\end{bmatrix} . \qquad (3.102)
$$

The matrix on the left hand side is called a *Sylvester matrix*. The system has a unique solution if (and only if) $\bar{A}$ and $\bar{B}$ have no common factors (that are not also factors of C). In this case the Sylvester matrix will be singular. The equations can be solved by using for example Gaussian elimination or LU decomposition (Press et al., 1988). Commonly, the time delay $d = 1$. Thus $r_0 = 1$ and the dimension of the system (3.102) can be reduced by removing this coefficient (Åström and Wittenmark, 1995). More details on Diophantine equations and their solution can be found in Åström and Wittenmark (1995), Mohtadi (1988), and Kučera et al. (1991).

Pole placement with all zeros canceled. If one prefers to have complete control over the closed-loop response, it is necessary to cancel the zeros of the system. This is done by including the zeros in the R-polynomial in addition to the integrator

$$R(q^{-1}) = \Delta B(q^{-1}) \bar{R}(q^{-1}) . \qquad (3.103)$$

If no additional zeros are introduced in B_m, the identity (3.92) degenerates to

$$q^{-d} \frac{B_m(q^{-1}) A_o(q^{-1})}{A_m(q^{-1}) A_o(q^{-1})} = q^{-d} \frac{T(q^{-1})}{\Delta A(q^{-1}) \bar{R}(q^{-1}) + q^{-d} S(q^{-1})} . \qquad (3.104)$$

There is in this case only a need for an observer polynomial of order $\deg(A_o) = \deg(A) - \deg(B) - 1$. For the typical sampled-data system with a time delay of one sampling period, $\deg(A) = \deg(B) - 1$. In this case no observer polynomial is required.

The controller polynomials are determined from the expressions (3.105) to (3.107):

$$T(q^{-1}) = A_m(q^{-1})A_o(q^{-1}) \tag{3.105}$$
$$A_m(q^{-1})A_o(q^{-1}) = \Delta A(q^{-1})\bar{R}(q^{-1}) + q^{-d}S(q^{-1}) \tag{3.106}$$
$$R(q^{-1}) = \bar{R}(q^{-1})\Delta B(q^{-1}) , \tag{3.107}$$

where
$$\deg(\bar{R}) = d - 1 \qquad \deg(S) = \deg(A) . \tag{3.108}$$

Pole placement with all zeros canceled is closely related to feedback linearization and control by inverse models treated earlier. Since the zeros, i.e., the roots of $B(q^{-1})$, enter the closed-loop characteristic polynomial they must lie inside the unit circle to ensure stability. Preferably they should furthermore be reasonably well-damped. In contrast to the previously discussed controllers it is in this case quite easy to check if the inverse of the closed-loop system is stable by simply calculating the roots of the extracted B-polynomials. If it turns out that certain zeros are outside the unit circle, it is necessary to leave the zeros unchanged instead of cancelling them.

Enforcing desired features on the controller. Suppose now that a disturbance, $v(t)$, is acting on system and let it be modelled as an additive disturbance entering the system at its input (see Figure 3.19).

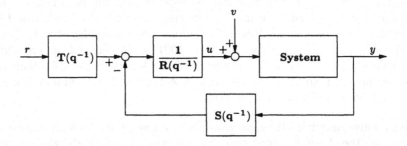

Figure 3.19. A disturbance acting on the input of the system.

If the system to be controlled is well-described by the linearized model, the closed-loop transfer function from disturbance to output is

$$y(t) = q^{-d}\frac{B(q^{-1})R(q^{-1})}{A(q^{-1})R(q^{-1}) + q^{-d}B(q^{-1})S(q^{-1})}v(t) \tag{3.109}$$

If there exists a polynomial in the delay operator, $\phi(q^{-1})$, such that

$$\phi(q^{-1})v(t) = 0 \tag{3.110}$$

it is clearly possible to compensate for the disturbance by including $\phi(q^{-1})$ as a factor of the controller polynomial $R(q^{-1})$. The way in which the integral action previously was enforced was actually a utilization of this principle:

$$v(t) = v_0 \quad \Rightarrow \quad \Delta v(t) = 0 . \tag{3.111}$$

The idea is to solve the Diophantine equation by assuming that ϕ is a factor of A and subsequently make ϕ a factor of R. This is shown below (for simplicity only the case where no zeros are canceled is shown here). To satisfy causality conditions the denominator of the desired closed-loop model, A_m, must be selected as a stable polynomial of degree

$$\deg(A_m) = \deg(A) + \deg(\phi) . \tag{3.112}$$

The polynomials R and S are then found by solving

$$A_m(q^{-1})A_o(q^{-1}) = A(q^{-1})\phi(q^{-1})\bar{R}(q^{-1}) + q^{-d}B(q^{-1})S(q^{-1}) \tag{3.113}$$
$$R(q^{-1}) = \phi(q^{-1})\bar{R}(q^{-1}) , \tag{3.114}$$

where

$$\begin{cases} \deg(S) = \deg(A) + \deg(\phi) - 1 \\ \deg(\bar{R}) = \deg(A) - 1 \quad \Rightarrow \quad \deg(R) = \deg(A) + \deg(\phi) - 1 . \end{cases} \tag{3.115}$$

It has already been shown how to compensate for constant disturbances. Sinusoidal disturbances are frequently encountered as well. This type of disturbance is easily compensated for too

$$v(t) = H \sin(\omega t T_s) \quad \Rightarrow \quad \left[1 - 2q^{-1}\cos(\omega T_s) + q^{-2}\right]v(t) = 0 . \tag{3.116}$$

Following this principle, it is straightforward to compensate for many disturbances with known characteristics. The method is sometimes referred to as the *internal model principle* because a model of the disturbance is included in the controller. Useful insights on the topic can be found in Åström and Wittenmark (1990) and Åström and Wittenmark (1995).

3.7.4 Pole Placement Control of the Benchmark System

The discussion of pole placement is concluded by testing the approximate pole placement design on the benchmark system. The same neural network model of the system as applied previously (e.g., in Section 3.2.4) is reused here.

Consider initially pole placement with full zero cancellation. Since in this case there is no need for an observer polynomial, the only design parameter is A_m. For comparison with the previous examples A_m is chosen to have two of its roots situated in $z = 0.7$ and one in $z = 0$

$$A_m(q^{-1}) = 1 - 1.4q^{-1} + 0.49q^{-2} + 0q^{-3} . \tag{3.117}$$

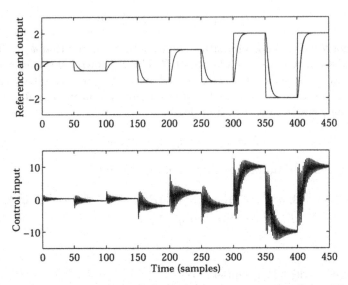

Figure 3.20. Control of the benchmark system using an approximate pole placement controller that cancels the zero. Upper: reference, system output, and output of desired closed-loop model. Lower: control signal.

The result of the simulation is displayed in Figure 3.20. The plot in the top panel compares reference, output of the system, and output of the desired closed-loop model. A perfect model-following behavior is achieved; it is impossible to distinguish the actual output of the system from the desired output. The controller performs similarly to the feedback linearizing controller in Section 3.4.3, which was also designed to place two poles in $z = 0.7$.

To assess the "degree of nonlinearity" it is informative to plot the coefficients of the extracted linear models. This provides a useful understanding of how the behavior of the system varies with the operating conditions. See Figure 3.21.

Alternatively one can calculate the poles and the zero of each of the extracted models and plot them in the complex plane. Such a *PZ map* is shown in Figure 3.22. It appears that the zero at all times is in the left half plane and is close to the unit circle. As discussed previously, this leads to a very oscillating control signal because the zero is made a factor of the R-polynomial. As a means to avoid this, consider now pole placement without zero cancellation. For causality reasons there is in this case a need for an observer polynomial of order 1. This is selected so that it has its root in $z = 0$

$$A_o(q^{-1}) = 1 - 0q^{-1} \,. \tag{3.118}$$

The result of the simulation is shown in Figure 3.23.

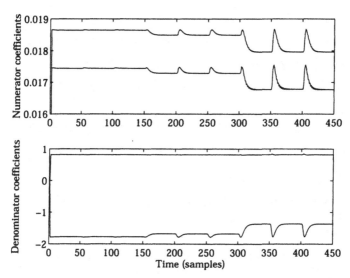

Figure 3.21. Coefficients of extracted linear models evolving with time. Upper panel: numerator coefficients. Lower panel: denominator coefficients.

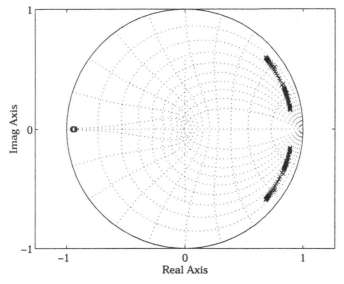

Figure 3.22. The two poles ('x') and the zero ('o') of the extracted linear models are here displayed in the complex plane.

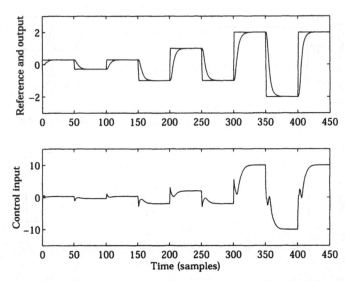

Figure 3.23. Control of the benchmark system using pole placement without zero cancellation.

Although the response essentially matches the one shown in Figure 3.20, it appears that a remedy for the problems with oscillating control signals has obviously been found.

The plot of the coefficients of the linearized models and the PZ map both showed how the linearization technique provides an excellent tool for analyzing the dynamics of a nonlinear system. By presenting the information in other ways an even better intuitive understanding of the system dynamics is possible. In conventional control design, for example, it is common practice to analyze second-order systems in terms of damping factor, ζ, and natural frequency, ω (Franklin et al., 1995):

$$H(s) = \frac{k}{s^2 + 2\zeta\omega s + \omega^2} \cdot \qquad (3.119)$$

After finding the continuous-time counterparts to the poles of the linearized models, damping and natural frequency are easily determined. Figure 3.21 shows how they evolved in the simulation without zero cancellation.

As the open-loop simulation in Section 3.1.1 also indicated, the natural frequency is increased and the damping reduced with the magnitude of the output.

3.7.5 Approximate Minimum Variance Design

As opposed to specifying desired behavior of the closed-loop system in terms of a transfer function or a set of closed-loop poles, one can specify the behavior

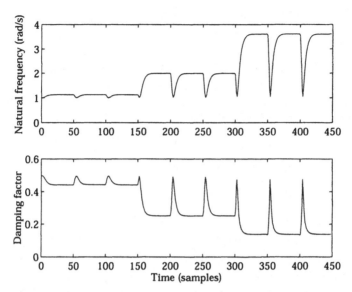

Figure 3.24. Natural frequency (top) and damping factor (bottom) of the linearized models.

in terms of a criterion which should be minimized. Criterion-based design has the advantage that tuning often becomes more intuitive, and moreover it is a quite natural way of designing controllers for stochastic systems. In the context of stochastic self-tuning regulators a popular class of controllers are the so-called *minimum variance controllers*. A few examples will be given in the following. A broader coverage can be found in Isermann et al. (1992). The results are presented for ARMAX models

$$A(q^{-1})y(t) = q^d B(q^{-1})u(t) + C(q^{-1})e(t) + \zeta(\tau) \qquad (3.120)$$

but are of course easily modified to conform to ARX descriptions as well by selecting $C(q^{-1}) = 1$. To obtain a controller with integral action, it is further assumed that the bias term, $\zeta(\tau)$, is modelled as integrated white noise. That is, the following ARIMAX (integrated ARMAX) type of model description is considered:

$$A(q^{-1})y(t) = q^d B(q^{-1})u(t) + \frac{C(q^{-1})}{\Delta}e(t) . \qquad (3.121)$$

The basic minimum variance controller is designed to solve the regulation problem. The objective is to compensate for stochastic disturbances and not to follow a reference trajectory. The criterion to be minimized is

$$J(t) = \mathbf{E}\left\{y^2(t)\right\} , \qquad (3.122)$$

where the minimizing control input is assumed to be a function of available observations and control inputs. That is,

$$u(t) = \arg\min_u \mathbf{E}\left\{y^2(t)|I_t\right\} , \tag{3.123}$$

where

$$I_t = \{y(t),\ y(t-1),\ \dots,y(0),\ u(t-1),\ \dots,u(0)\} ,$$

The controller polynomials, R, S, are found as the solution to (Isermann et al., 1992; Åström and Wittenmark, 1995):

$$C(q^{-1}) = \Delta A(q^{-1})\bar{R}(q^{-1}) + q^{-d}S(q^{-1}) \tag{3.124}$$
$$R(q^{-1}) = \bar{R}(q^{-1})\Delta B(q^{-1}) \tag{3.125}$$

where

$$\deg(\bar{R}) = d - 1 \qquad \deg(S) = \max(n_a, n_c - d) .$$

The T polynomial is irrelevant in this case since there is no reference signal.

One can interpret the minimum variance controller as a pole placement controller with all zeros canceled and with $B(q^{-1})C(q^{-1})$ as the characteristic polynomial. The design is consequently restricted to systems with a stable inverse (all zeros inside the unit circle). In practice the roots of $B(q^{-1})$ should preferably not be too close to the unit circle. In principle $C(q^{-1})$ should automatically have its roots inside the unit circle. However, due to the linearization or to possible modelling errors it may happen that $C(q^{-1})$ in some regimes will have its roots outside the unit circle. To overcome this problem, one can use *spectral factorization* to "project" the unstable roots back inside the stability domain (Åström and Wittenmark, 1995):

As it is assumed that the noise acting on the input of the system is a stationary stochastic process, the spectral density exists and is given by

$$\Phi(\omega) = \frac{1}{2\pi}C\left(e^{i\omega}\right)C\left(e^{-i\omega}\right)\sigma_e^2 . \tag{3.126}$$

Notice that the C polynomial can also be factorized as

$$\begin{aligned}
C(z) &= 1 + c_1 z^{-1} + c_2 z^{-1} + \cdots + c_k z^{-k} \\
&= (1 - p_1 z^{-1})(1 - p_2 z^{-1})\cdots(1 - p_k z^{-1}) \\
&= (1 - e^{p_1' T_s} z^{-1})(1 - e^{p_2' T_s} z^{-1})\cdots(1 - e^{p_k' T_s} z^{-1}) .
\end{aligned} \tag{3.127}$$

If somehow the C polynomial obtained by linearization has roots outside the unit circle, it can be replaced by a new polynomial, C^*, that corresponds to C except that roots outside the unit circle, $z = e^{ip_i'} = p_i$, are replaced by roots inside: $z = e^{-ip_i'} = 1/p_i$. It is clear that substitution of C^* for C in (3.126) will not affect the calculation of the spectral density:

$$2\pi\Phi(\omega) = C\left(e^{i\omega}\right)C\left(e^{-i\omega}\right)\sigma_e^2 = C^*\left(e^{i\omega}\right)C^*\left(e^{-i\omega}\right)\sigma_e^2 . \tag{3.128}$$

To apply the minimum variance philosophy to a more wide-ranging class of systems and applications, many different extensions have been developed. To unify the different variations on minimum variance type control the so-called *Generalized Minimum Variance controller (GMV)* has been proposed (Clarke and Gawthrop, 1975; Gawthrop, 1977). The GMV criterion takes the form

$$J(t) = \mathbf{E} \left\{ \left[P(q^{-1})y(t+d) - W(q^{-1})r(t) \right]^2 + \left[Q(q^{-1})u(t) \right]^2 \middle| I_t \right\} , \quad (3.129)$$

where P, W and Q are rational transfer functions. Clearly this controller solves the servo problem rather than only the regulation problem. Moreover, systems with an unstable inverse can be controlled with a sensible choice of design parameters.

GMV is obviously a fairly general approach with many design parameters. In most practical situations one will start out with a simplified criterion as this will facilitate tuning and implementation. A common choice is the MV1-criterion (Isermann et al., 1992):

$$J(t) = \mathbf{E} \left\{ \left[y(t+d) - r(t) \right]^2 + \rho \left[\Delta u(t) \right]^2 \middle| I_t \right\} . \quad (3.130)$$

The minimizing solution for this is

$$R(q^{-1}) = \Delta B(q^{-1}) \bar{R}(q^{-1}) + \rho \Delta C(q^{-1}) \quad (3.131)$$
$$T(q^{-1}) = C(q^{-1}) . \quad (3.132)$$

$\bar{R}(q^{-1})$ and $S(q^{-1})$ are found from the Diophantine equation

$$C(q^{-1}) = \Delta A(q^{-1}) \bar{R}(q^{-1}) + q^{-d} S(q^{-1}) , \quad (3.133)$$

where

$$\deg(\bar{R}) = d - 1 \qquad \deg(S) = \max(n_a, n_c - d)$$

From the expression (3.131) the reason for penalizing the square of *differenced* control inputs is evident. The differencing is necessary in order to accomplish the desired integral action; that is, Δ becomes a factor of R.

An alternative criterion-based design called *predictive control* will be examined in the following section (Section 3.8). The reason for devoting a separate section to this design is that a solution can be found not only for models achieved through instantaneous linearization but for full non-linear neural network models as well.

3.7.6 Section Summary

Prior to this section only *direct designs* had been presented. In this section the first examples of *indirect designs* were given. That is, the controller was not itself a neural network, but its design was based on a neural network model of the system to be controlled. A method was proposed for linearizing a neural network model around its current point of operation, and it was discussed how this *instantaneous linearization technique* could be used for designing a gain scheduling type of control system. It was shown how the linearized models could be used for deriving approximate pole placement and minimum variance type controllers. It was also discussed how the linearized models could provide a useful interpretation of the system dynamics (poles, zeros, damping, natural frequency, etc.).

By applying instantaneous linearization in the design of controllers for nonlinear systems it appeared that a number of serious drawbacks were overcome. However, the design should be used with some care. If the system contains hard nonlinearities in certain regimes the linearization might offer a very poor description here. It is easy to imagine a situation where this could have a serious impact on the controller design.

Below the main characteristics are summarized:

Advantages:

- Possible to use well-established linear design techniques.

- Implementation reasonably simple.

- Well-suited for real-time use. Linearization+controller design can be carried out in between samples.

- Systems with an unstable inverse can be controlled if the zeros are not canceled.

- Interpretation of the behavior of the closed-loop system and the controller is quite easy.

- Provides a useful physical interpretation of the dynamics of the system.

Disadvantages:

- The linearized model might be valid only in a very narrow region around the current operating point.

3.8 Predictive Control

Criterion-based designs have a major advantage as far as tuning is concerned. Selecting weight factors are generally more intuitive than specifying a desired closed-loop model. The optimal control method was proposed as a simple method for training a neural network as a detuned inverse of the system. Unfortunately, it had the drawback of being a direct adaptive control strategy in the sense that the controller network was trained on-line to minimize the criterion. Not only can this cause stability problems when in certain regimes the weights have not converged adequately, it also makes tuning time-consuming because the controller must be retrained each time the criterion is modified. The instantaneous linearization technique opened up for all the well-known linear design techniques, such as minimum variance and linear quadratic regulators. This technique is thus extremely flexible in terms of possible design methods. However, it possesses the weakness of relying on approximate linear models. In particular for systems with hard nonlinearities, i.e., when the linear model in certain regimes is valid only in a very narrow region around the operating point, this may be a problem. A flexible and very powerful criterion-based design, which does not suffer from these drawbacks, is the so-called *predictive controller*.

Predictive control can be used together with instantaneous linearization in an implementation similar to the approximate minimum variance design. However it applies to a full nonlinear model of the system as well. Due to its flexibility, predictive control can handle a large class of systems and despite the fact there are many design parameters, the controller can be tuned by quite intuitive means. The type of predictive control examined here is usually referred to as *Generalized Predictive Control* or *GPC*. It was originally proposed in Clarke et al. (1987a) and Clarke et al. (1987b) as an alternative to pole placement and minimum variance designs used in self-tuning regulators. The argument for introducing GPC in a self-tuning context was that it was based on a more flexible criterion than (generalized) minimum variance controllers without requiring an excessive amount of computations. Although it originated in an adaptive control context, GPC has many attractive features which definitely makes it worthwhile considering even for time-invariant systems.

The idea behind generalized predictive control is at each iteration to minimize a criterion of the following type

$$J(t, U(t)) = \sum_{i=N_1}^{N_2} [w(t+i) - \hat{y}(t+i)]^2 + \rho \sum_{i=1}^{N_u} [\Delta u(t+i-1)]^2 \quad (3.134)$$

with respect to the N_u future control inputs

$$U(t) = [u(t) \ ... \ u(t+N_u-1)]^T \quad (3.135)$$

and subject to the constraint

$$\Delta u(t+i) = 0, \qquad N_u \leq i \leq N_2 - d . \quad (3.136)$$

N_1 is denotes the *minimum prediction (or cost) horizon*, N_2 denotes the *prediction (or maximum cost) horizon*, and N_u the *(maximum) control horizon*. ρ is a weighting factor penalizing changes in the control input.

For nonlinear systems the optimization problem must be solved at each sample, resulting in a sequence of future control inputs. From this sequence, the first component, $u(t)$, is then applied to the system. For linear systems the optimization problem can be solved beforehand and the solution transformed into the RST-controller structure examined in Section 3.7. The predictive controller principle is depicted in Figure 3.25.

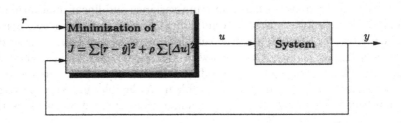

Figure 3.25. The principle of predictive control.

A prime characteristic that distinguishes predictive control from other controllers, is the idea of a *receding horizon*; at each sample the control signal is determined to achieve a desired behavior in the following N_2 time steps. This idea is also appealing because it relates to many of the control tasks that one as a human being carries out on a daily basis. This intuitive foundation can to some extent accommodate the tuning of the design parameters.

Another important property is the notion of a control horizon, which is smaller than the prediction horizon. The principle is that only the first N_u future control inputs are determined. From that point on, the control input is assumed constant. The reason for introducing a control horizon is that the computational burden decreases significantly as the length of this horizon is shortened. As a consequence, the horizon is typically kept as short as possible. In fact, to allow for a reasonably rapid sampling frequency, $N_u = 1$ is used in most applications.

In the original presentation of GPC (Clarke et al., 1987a,b), $\hat{y}(t+k)$ represented the minimum variance k-step ahead predictions. However, as discussed in Soeterboek (1992) one might prefer a more pragmatic approach and simply treat the predictor as an additional design parameter (in the sense that one uses a predictor with other properties than to be minimum variance). This section embraces two different types of predictors:

- Successive recursion of a deterministic neural network model.

- An approximate minimum variance predictor based on instantaneous linearization of an NNARX model.

In the first case, the predictor is nonlinear in the future control inputs; thus the minimization task has to be performed with an iterative search method. In the second case a unique solution exists, and the future control inputs can be found directly.

First the "true" nonlinear predictive controller (NPC) will be described. The focus will be on derivation of the most suitable optimization algorithms for minimization of the GPC criterion. The approximate predictive controller (APC), which is based on the instantaneous linearization is introduced next. Finally, a few comments will be given on how to extend the basic criterion to account for actuator limitations or compensate for disturbances with known characteristics.

3.8.1 Nonlinear Predictive Control (NPC)

Minimization of the GPC criterion, when the predictions are determined by a nonlinear relation of the future inputs, constitutes a complex nonlinear programming problem. Unfortunately, the problem does not exactly become less involved as the real-time implementation issues are taken into account. These demand that a limited maximum response time exists (which preferably is short) and that the control law is numerically robust and does not have problems with convergence. Moreover, it is instrumental that the minimization can run automatically since there will be no possibility for user interference in real time. Consequently, it cannot depend on numerous design parameters which need proper adjustment to achieve a satisfactory convergence. Below, two different algorithms will be described, which are considered suitable for the task and various issues relating to the implementation and controller tuning will be addressed.

The Criterion. To alleviate the derivation of the control law, the criterion is rewritten in vector notation as follows

$$J(t, U(t)) = \left[R(t) - \hat{Y}(t) \right]^T \left[R(t) - \hat{Y}(t) \right] + \rho \tilde{U}^T(t) \tilde{U}(t) \quad (3.137)$$
$$= E^T(t) E(t) + \rho \tilde{U}^T(t) \tilde{U}(t),$$

where

$$\begin{aligned}
R(t) &= [r(t + N_1) \; ... \; r(t + N_2)]^T \\
\hat{Y}(t) &= [\hat{y}(t + N_1|t) \; ... \; \hat{y}(t + N_2|t)]^T \\
E(t) &= [e(t + N_1|t) \; ... \; e(t + N_2|t)]^T \\
\tilde{U}(t) &= [\Delta u(t) \; ... \; \Delta u(t + N_u - 1)]^T
\end{aligned} \qquad (3.138)$$

and

$$e(t + k|t) = r(t + k) - \hat{y}(t + k|t) \quad \text{for} \quad k = N_1, \; ... \; , N_2 . \qquad (3.139)$$

The Predictor. Provided that the system to be controlled is deterministic, the one-step ahead prediction is given by

$$\hat{y}(t) = \hat{y}(t|t-1) = \hat{g}_1[y(t-1), \ ... \ , y(t-n),$$
$$u(t-d), \ ... \ , u(t-d-m)] \ . \qquad (3.140)$$

g is some function realized by a neural network and d is the time delay, which is assumed to be at least one. The k-step ahead prediction of the system's output can then be calculated by shifting the expression forward in time while substituting predictions for actual measurements where these do not exist. This can also be written:

$$\hat{y}(t+k) = \hat{y}(t+k|t)$$
$$= \hat{g}[\hat{y}(t+k-1), \ ... \ , \hat{y}(t+k-\min[k,n]),$$
$$y(t-1), \ ... \ , y(t-\max[n-k,0]),$$
$$u(t+k-d), \ ... \ , u(t+k-d-m)] \ . \qquad (3.141)$$

It is assumed that the observation of the output is available up to time $t-1$, only. For this reason $\hat{y}(t)$ enters the expression instead of the real output $y(t)$. This is to avoid the computational time delay, which would appear if $u(t)$ was to be calculated between D/A and A/D-converter calls at time t. Instead $u(t)$ can be calculated after the D/A-call in the previous sample. This same assumption was sometimes made in early state space control design, in which case the states were estimated with a *predictive observer* based only on past observations (Kwakernaak and Sivan, 1972).

By inserting g, which here is assumed to be a two-layer MLP-network with *tanh* activation functions in the hidden units and a linear output unit, one obtains

$$\hat{y}(t+k) = \sum_{j=1}^{n_h} W_j f\left(\tilde{h}(h,j)\right) + W_0 \ . \qquad (3.142)$$

$f(x) = \tanh(x)$ and

$$\tilde{h}(k,j) = \sum_{i=1}^{\min(k,n)} w_{j,i}\hat{y}(t+k-i) \ + \sum_{i=\min(k,n)+1}^{n} w_{j,i}y(t+k-i)$$

$$+ \sum_{i=0}^{m} w_{j,n+i+1}u(t+k-d-i) \ + \ w_{j,0} \ . \qquad (3.143)$$

It is also possible to train $N_2 - N_1 + 1$ networks to directly produce each of the predictions $\hat{y}(t+k|t)$, $N_1 \leq k \leq N_2$. However, this method has certain deficiencies as more distant predictions are considered. The networks require an additional input each time k is increased. To obtain reasonably accurate long range predictions, the condition of proper excitation thus demands gigantic training sets. For this reason, only predictions determined in the recursive fashion described above will be considered here.

Deriving the Control Law. Minimizing the GPC criterion when the predictions are nonlinear in the control inputs is a quite elaborate optimization problem. In order to determine the minimum it is necessary to apply an iterative search method similar to the strategies used when training neural networks

$$U^{(i+1)} = U^{(i)} + \mu^{(i)} f^{(i)} . \tag{3.144}$$

$U^{(i)}$ specifies the current iterate of the sequence of future control inputs, $\mu^{(i)}$ specifies the step size, and $f^{(i)}$ the search direction.

As for the network training problem, numerous choices of search direction and step size may come into consideration. However, several issues distinguish the present problem from network training:

- Most often fast convergence is an absolute necessity.

- Minimum must be found in real-time; i.e., a maximum response time must exist.

- Numerical robustness is crucial.

- Not a mean-square-error type criterion.

- Only a few parameters need to be determined since the control horizon, N_u, usually is small.

- The accuracy is often not of particular importance. An accuracy better than the resolution of the D/A-converter is meaningless.

These aspects govern a somewhat different approach to the optimization problem. A pure gradient method will in general have problems satisfying the first two demands; thus, only second-order methods are considered. First a review of the options.

The Gauss-Newton and Levenberg-Marquardt Approaches. One can in principle proceed as in Section 2.4 by developing an algorithm-based on a first-order approximation of the error between reference and output, $e(t + k)$. Unfortunately a rapid convergence cannot be expected with the method as generally it is not a *small residual problem* (Dennis and Schnabel, 1983). Due to the weighting of the (differenced) control inputs, the errors, $e(t + k)$ will often be significant.

The Newton-based Levenberg-Marquardt Approach. Alternatively the Levenberg-Marquardt modification employed to the full Newton method is considered. The search direction is in this case found from

$$\left(\tilde{H} \left[U^{(i)}(t) \right] + \lambda^{(i)} I \right) f^{(i)} = -G \left[U^{(i)}(t) \right] . \tag{3.145}$$

Gradient vector and Hessian matrix are found by

$$G\left[U^{(i)}(t)\right] = \left.\frac{\partial J(t, U(t))}{\partial U(t)}\right|_{U(t)=U^{(i)}(t)} \tag{3.146}$$

$$= \left.-2\Phi^T\left[U^{(i)}(t)\right]\tilde{E}(t) + 2\rho\frac{\partial \tilde{U}^T(t)}{\partial U(t)}\tilde{U}(t)\right|_{U(t)=U^{(i)}(t)}$$

$$H\left[U^{(i)}(t)\right] = \left.\frac{\partial^2 J(t, \bar{U}(t))}{\partial U(t)^2}\right|_{U(t)=U^{(i)}(t)} \tag{3.147}$$

$$= \left.\frac{\partial}{\partial U(t)}\left(\frac{\partial \hat{Y}^T(t)}{\partial U(t)}E(t)\right) + 2\rho\frac{\partial \tilde{U}^T(t)}{\partial U(t)}\frac{\partial \tilde{U}(t)}{\partial U(t)}\right|_{U(t)=U^{(i)}(t)}$$

The second term of the Hessian is clearly positive semidefinite, but this is not necessarily the case for the first term. Occasionally this may become non-positive definite. Selecting a sufficiently large weighting factor, ρ, can indeed ensure that the Hessian stays positive definite, but of course it is not a feasible strategy in practice. A small modification of the Levenberg-Marquardt algorithm will resolve this problem. An algorithm will be presented later in this section.

As one will soon see, the expressions for gradient and Hessian are quite complex and an extensive amount of computations is required. As a consequence one might consider approximating the quantities numerically rather than by using analytical expressions (Pröll and Karim, 1994). While the accuracy of the Hessian typically is of minor importance, using approximations of the gradient is in general dissuaded when it is feasible to calculate it by analytical means, though. The reason for this being that the true gradient is vital for pointing out valid descent directions (Dennis and Schnabel, 1983).

The Quasi-Newton Approach. A third method is to directly construct a positive definite approximation of the inverse Hessian matrix through application of the so-called Broyden-Fletcher-Goldfarb-Shanno ("BFGS") algorithm (Dennis and Schnabel, 1983). This algorithm leads to a positive definite approximation of the inverse Hessian using the information embedded in previous evaluations of gradient and criterion. The algorithm has good local convergence properties and is popular when the Hessian is difficult derive by analytical means or time-consuming to compute. The search directions is

$$f^{(i)} = -B^{(i)}G\left[U^{(i)}(t)\right], \tag{3.148}$$

where $B^{(i)}$ specifies the approximation of the inverse Hessian. In the Levenberg-Marquardt method the step size was controlled along with the search direction. The Quasi-Newton method approximates the full Newton search direction and it is therefore necessary to complement it with a line search to ensure convergence. There are quite restrictive rules for how this line search is implemented to guarantee the validity of the BFGS-update. A complete Quasi-Newton algorithm is suggested later in this section.

Computing the Partial Derivatives. Before calculating the search directions, it is necessary to determine the various partial derivatives entering the different expressions. These are derived in the following.

The partial derivative $\dfrac{\partial \tilde{U}(t)}{\partial U(t)}$:

Since $\Delta u(t) = u(t) - u(t-1)$ this is

$$
\frac{\partial \tilde{U}(t)}{\partial U(t)} =
\begin{bmatrix}
1 & 0 & 0 & 0 & 0 \\
-1 & 1 & 0 & 0 & 0 \\
\vdots & \ddots & \ddots & 0 & \vdots \\
0 & 0 & -1 & 1 & 0 \\
0 & 0 & 0 & -1 & 1
\end{bmatrix},
\tag{3.149}
$$

which is a matrix of dimension $N_u \times N_u$. The derivative is independent of time and can be constructed beforehand.

The partial derivative $\dfrac{\partial \hat{Y}(t)}{\partial U(t)}$:

This is a matrix of dimension $N_u \times (N_2 - N_1 + 1)$

$$
\frac{\partial \hat{Y}(t)}{\partial U(t)} =
\begin{bmatrix}
\frac{\partial \hat{y}(t+N_1)}{\partial u(t)} & \cdots & \frac{\partial \hat{y}(t+N_1)}{\partial u(t+N_u-1)} \\
\vdots & \ddots & \vdots \\
\frac{\partial \hat{y}(t+N_2)}{\partial u(t)} & \cdots & \frac{\partial \hat{y}(t+N_2)}{\partial u(t+N_u-1)}
\end{bmatrix}.
\tag{3.150}
$$

To accommodate the calculation of the derivatives of which this matrix is composed, the following separation between terms depending on past and future control inputs is made

$$
\begin{aligned}
\tilde{h}(k,j) =\ & \sum_{i=1}^{\min(k-d,n)} w_{j,i}\hat{y}(t+k-i) \\
&+ \sum_{i=1}^{\min(k-d-N_u+2,m+1)} w_{j,n+i}u(t+N_u-1) \\
&+ \sum_{i=k-d-N_u+2}^{\min(k-d,m)} w_{j,n+i+1}u(t-d+k-i) \\
&+ \sum_{i=k-d+1}^{\min(k,n)} w_{j,i}\hat{y}(t+k-i) \\
&+ \sum_{i=k+1}^{n} w_{j,i}y(t+k-i) \\
&+ \sum_{i=k-d+1}^{m} w_{j,n+i+1}u(t-d+k-i)\ +\ w_{j,0}.
\end{aligned}
\tag{3.151}
$$

The first three sums depend on future control inputs while the remaining three depend on past control inputs only.

For all $k \in [d, N_2]$ and for all $l \in [0, \min(k-1, N_u - 1)]$ determine

$$\frac{\partial \hat{y}(t+k)}{\partial u(t+l)} = \sum_{j=1}^{n_h} W_j f' \left[\tilde{h}(k,j) \right] \frac{\partial \tilde{h}(k,j)}{\partial u(t+l)}$$

$$= \sum_{j=1}^{n_h} W_j f' \left[\tilde{h}(k,j) \right] h(k,l,j) , \qquad (3.152)$$

where

$$h(k,l,j) = \sum_{i=1}^{\min(k-d,n)} w_{j,i} \frac{\partial \hat{y}(t+k-i)}{\partial u(t+l)}$$

$$+ \sum_{i=1}^{\min(k-d-N_u+2,m+1)} w_{j,n+i} \frac{\partial u(t+N_u-1)}{\partial u(t+l)}$$

$$+ \sum_{i=k-d-N_u+2}^{\min(k-d,m)} w_{j,n+i+1} \frac{\partial u(t-d+k-i)}{\partial u(t+l)} . \qquad (3.153)$$

Since

$$\frac{\partial u(t+N_u-1)}{\partial u(t+l)} = \begin{cases} 1, & l = N_u - 1 \\ 0, & \text{otherwise} \end{cases} \qquad (3.154)$$

$$\frac{\partial u(t-d+k-i)}{\partial u(t+l)} = \begin{cases} 1, & l = k-d-i \\ 0, & \text{otherwise} \end{cases} \qquad (3.155)$$

$$\frac{\partial \hat{y}(t+k-i)}{\partial i + l)} = 0, \qquad l \geq k-d-i+1 \qquad (3.156)$$

the expression for $h(k,l,j)$ can be reduced to

$$h(k,l,j) = \sum_{i=1}^{min(k-d-l,n)} w_{j,i} \frac{\partial \hat{y}(t+k-i)}{\partial u(i+l)}$$

$$+ \sum_{i=1}^{min(k-d-N_u+2,m+1)} w_{j,n+i} , \qquad l = N_u - 1 \qquad (3.157)$$

$$h(k,l,j) = \sum_{i=1}^{min(k-d-l,n)} w_{j,i} \frac{\partial \hat{y}(t+k-i)}{\partial u(i+l)} + w_{j,n+k-d-l+1} ,$$

$$\max(0, k-d-m) \leq l \leq N_u - 2 \qquad (3.158)$$

$$h(k,l,j) = \sum_{i=1}^{min(k-d-l,n)} w_{j,i} \frac{\partial \hat{y}(t+k-i)}{\partial u(i+l)} ,$$

$$0 \leq l < \max(0, k-d-m) . \qquad (3.159)$$

When inserting the hidden layer activation functions it is recalled that

$$f(x) = \tanh(x) \;\Rightarrow\; f'(x) = 1 - f^2(x) \,. \tag{3.160}$$

The order in which the derivatives in (3.152) are calculated is very important since the calculations should consist of known quantities only. The procedure oulined in Table 3.3 should be followed.

Table 3.3. Procedure for calculating the entries of the matrix defined in (3.150).

---
for $k = d$ to N_2
  for $l = 0$ to $\min(k - 1, N_u - 1)$
    calculate $\frac{\partial \hat{y}(t+k)}{\partial u(t+l)}$
  end
  {The following should only be performed once}
  for $l = k$ to $N_u - 1$
    set $\frac{\partial \hat{y}(t+k)}{\partial u(t+l)} = 0$
  end
end

The partial derivative $\dfrac{\partial}{\partial U(t)}\left(\dfrac{\partial \hat{Y}^T(t)}{\partial U(t)} E(t)\right)$:

The vector inside the parenthesis is

$$\frac{\partial \hat{Y}^T(t)}{\partial U(t)} E(t) =$$

$$\begin{bmatrix} \frac{\partial \hat{y}(t+N_1)}{\partial u(t)} e(t + N_1) + \cdots + \frac{\partial \hat{y}(t+N_2)}{\partial u(t)} e(t + N_2) \\ \vdots \\ \frac{\partial \hat{y}(t+N_1)}{\partial u(t+N_u-1)} e(t + N_1) + \cdots + \frac{\partial \hat{y}(t+N_2)}{\partial u(t+N_u-1)e(t+N_2)} \end{bmatrix} . \tag{3.161}$$

Taking the derivative of this with respect to $U(t)$ results in the following $N_u \times N_u$ matrix

$$\frac{\partial}{\partial U(t)}\left(\frac{\partial \hat{Y}^T(t)}{\partial U(t)} E(t)\right) = \begin{bmatrix} \frac{\partial^2 \hat{Y}^T(t)}{\partial u(t)^2} E(t) & \cdots & \frac{\partial^2 \hat{Y}^T(t)}{\partial u(t)\partial u(t+N_u-1)} E(t) \\ \vdots & \ddots & \vdots \\ \frac{\partial^2 \hat{Y}^T(t)}{\partial u(t)\partial u(t+N_u-1)} E(t) & \cdots & \frac{\partial^2 \hat{Y}^T(t)}{\partial u(t+N_u-1)^2} E(t) \end{bmatrix}$$

$$- \left[\frac{\partial \hat{Y}^T(t)}{\partial U(t)}\right]\left[\frac{\partial \hat{Y}(t)}{\partial U(t)}\right] . \tag{3.162}$$

The second term is formed by multiplying the vectors derived earlier. Calculation of the first term is more involved. To determine this, each of the second-order derivatives

$$\frac{\partial^2 \hat{y}(t+k)}{\partial u(t+l)\partial u(t+p)} = \tag{3.163}$$

$$\sum_{j=1}^{n_h} W_j \left(f'' \left[\tilde{h}(k,j) \right] h(k,p,j)h(k,l,j) + f' \left[\tilde{h}(k,j) \right] \frac{\partial h(k,l,j)}{\partial u(t+p)} \right)$$

where

$$\frac{\partial h(k,l,j)}{\partial u(t+p)} = \sum_{i=1}^{\min(k-d-l,n)} w_{j,i} \frac{\partial^2 \hat{y}(t+k-i)}{\partial u(t+l)\partial u(t+p)} \tag{3.164}$$

and

$$f(x) = \tanh(x) \quad \Rightarrow \quad f''(x) = -2f(x)f'(x) = -2f(x)\left[1 - f^2(x)\right] \tag{3.165}$$

need to be calculated for all $k \in [d, N_2]$, for all $l \in [0, \min(k-1, N_u-1)]$ and for all $p \in [0, l]$. The derivatives should be calculated according to the order specified in Table 3.4.

Table 3.4. Algorithm for calculation of the second-order derivatives.

> **for** $k=d$ **to** N_2
>   **for** $l = 0$ **to** $\min(k - 1, N_u - 1)$
>   **for** $p = 0$ **to** l
>   calculate $\frac{\partial \hat{y}^2(t+k)}{\partial u(t+l)\partial u(t+p)}$
>   **end**
>   **end**
>   {The following should only be performed once}
>   **for** $l = k$ **to** $N_u - 1$
>   **for** $p = 0$ **to** l
>   set $\frac{\partial \hat{y}^2(t+k)}{\partial u(t+l)\partial u(t+p)} = 0$
>   **end**
>   **end**
> **end**

A Full Newton-based Levenberg-Marquardt Algorithm. It has now been shown how to calculate all components essential for implementation of Newton-type algorithms. First a version of the Levenberg-Marquardt algorithm is proposed. The algorithm is different from the training algorithm discussed in Section 2.4 in that it is founded on a full Newton algorithm. The problem with the Newton algorithm is that the Hessian matrix is not guaranteed to be positive definite in an open neighborhood of a minimum. Fortunately is easily handled within the Levenberg-Marquardt framework (Fletcher, 1987).

The search direction for a Newton-based Levenberg-Marquardt algorithm was given by (3.145). Before this can be applied for updating the sequence of control inputs it is necessary to investigate whether this direction is a descent direction at all. This in turn implies that a definiteness check of the matrix $H\left[U^{(i)}(t)\right] + \lambda^{(i)}I$. If the matrix is not positive definite, the Levenberg-Marquardt parameter, λ, should be increased until this occurs. In Section 2.4 it was suggested that Cholesky factorization was deployed for solving the system of equations to find the search direction. As it turns out, Cholesky factorization is also an efficient method for determining whether or not a symmetric matrix is positive definite. In this algorithm Cholesky factorization is therefore used for determining whether or not positive definiteness is achieved. If not, λ is increased until it is. Next, the Cholesky factors are reused to calculate the search direction. The complete algorithm is presented in Table 3.5.

Since this has to work in real-time it is important that a (known) finite maximum response time exists. Unfortunately it is possible to run around in either of the two loops: *2-2* or *1-2-3-4-6-8-1* arbitrarily many times. Consequently it is necessary to impose a maximum limit on the number of times *Step 2* is executed.

A Quasi-Newton Algorithm. Apart from being difficult to implement, the weakness of a full Newton-type method is that from a computational perspective it becomes still less manageable as the prediction and control horizons are increased. The computational burden is due to the calculation of the gradient, Hessian, and the Cholesky factorization. An alternative, which to some extent remedies this computational overhead, is the so-called Quasi-Newton algorithm. In this algorithm the inverse Hessian is being approximated directly from the information about the criterion and gradient contained in the past iterations. Different update formulas for approximating the inverse Hessian exist. The most popular one, which is also the one used here, is the BFGS algorithm (BFGS=Broyden-Fletcher-Goldfarb-Shanno). See for example Dennis and Schnabel (1983).

In addition to the approximation of the inverse Hessian, the Quasi-Newton algorithm must also be accompanied by a line search for determination of the step size. This can be carried out in different ways but due to the nature of the present problem it is important that the search requires a minimum of criterion evaluations. In particular when the prediction horizon is long, evaluation of the criterion becomes very expensive. Two algorithms are pre-

Table 3.5. A full Newton-based Levenberg-Marquardt algorithm.

0: Initialize λ. Select initial sequence of future control inputs, $U^{(0)}$ and evaluate the criterion $J(U^{(0)})$. $i = 0$.

1: Compute gradient vector $G(U^{(i)})$ and Hessian matrix $H(U^{(i)})$.

2: Use Cholesky factorization on the matrix $H(U^{(i)}) + \lambda I$. If the matrix is not positive definite, the factorization will fail. Set $\lambda = 4\lambda$ and go to 2.

3: Determine the search direction, $f^{(i)}$, (use the Cholesky factors):

$$\left(H\left[U^{(i)} \right] + \lambda I \right) f^{(i)} = -G\left[U^{(i)} \right]$$

4: Evaluate the criterion $J(U^{(i)} + f^{(i)})$ and determine the ratio between actual and predicted decrease in the criterion:

$$r^{(i)} = 2\frac{J(U^{(i)}) - J(U^{(i)} + f^{(i)})}{\lambda \left(f^{(i)} \right)^T f^{(i)} - \left(f^{(i)} \right)^T G\left[U^{(i)} \right]}.$$

5: If $r^{(i)} > 0.75$ then $\lambda = \lambda/2$ and go to 7.

6: If $r^{(i)} < 0.25$ then $\lambda = 2\lambda$.

7: If $r^{(i)} > 0$ then: $U^{(i+1)} = U^{(i)} + f^{(i)}$, $i = i + 1$.

8: Go to 1 if stopping criterion: $|U^{(i)} - U^{(i-1)}| < \delta$ or $i >$(maximum # of iterations) is not satisfied.

9: Accept sequence of future control inputs and terminate.

---

sented in Table 3.6 and Table 3.7. The basic BFGS algorithm is outlined in Table 3.6 while the soft line search algorithm is given in Table 3.7.

While the basic algorithm is straightforward, the line search is somewhat involved. The primary objective of the line search is to ensure that the criterion is decreased from one iteration to the next, but it serves an additional purpose in this case. Provided that $B^{(i-1)}$ is positive definite, the BFGS update of $B^{(i)}$ is only guaranteed to be positive definite if the condition: $\Delta U^{(i)T} \Delta G^{(i)} > 0$ is satisfied. The line search must therefore also take care of this. An "exact" line search will do this, but it requires far too many criterion evaluations. Instead it is recommended to use a "soft" line search. The algorithm described in Table 3.6 corresponds to the one found in Madsen (1984). It is based on the ideas outlined in Dennis and More (1977), and the interested reader is referred to this reference for a description of the principle.

In the algorithm for selecting the step size (see Table 3.7), the following two conditions play a substantial role in the algorithm.

$$J(U^{(i)} + \mu f^{(i)}) \le J(U^{(i)}) + \delta \mu f^{(i)T} G\left[U^{(i)} \right] \qquad (3.166)$$

Table 3.6. Basic BFGS algorithm.

0: Select initial sequence of future control inputs. Evaluate criterion $J(t, U^{(0)})$ and gradient $G\left[U^{(0)}\right]$. Initialize the approximation to the inverse Hessian $B^{(0)} = I$. $i = 0$.

1: Determine search direction: $f^{(i)} = -B^{(i)}G\left[U^{(i)}\right]$.

2: Determine step size $\mu^{(i)}$ by "soft" line search. See the algorithm in Table 3.7.

3: Update sequence of future control inputs: $U^{(i+1)} = U^{(i)} + \mu^{(i)}f^{(i)}$.

4: Go to 8 if one if the stopping criteria: $|U^{(i)} - U^{(i-1)}| < \delta$ or $i >$(maximum # of iterations) are satisfied.

5: $i = i + 1$.

6: Update approximation to the inverse Hessian with the BFGS algorithm:

$$B^{(i)} = \left[I - \frac{\Delta U^{(i)}\left(\Delta G^{(i)}\right)^T}{\left(\Delta G^{(i)}\right)^T \Delta U^{(i)}}\right] B^{(i)} \left[I - \frac{\Delta G^{(i)}\left(\Delta U^{(i)}\right)^T}{\left(\Delta G^{(i)}\right)^T \Delta U^{(i)}}\right]$$

$$+ \frac{\Delta U^{(i)}\left(\Delta U^{(i)}\right)^T}{\left(\Delta G^{(i)}\right)^T \Delta U^{(i)}},$$

where

$$\Delta U^{(i)} \triangleq U^{(i)} - U^{(i-1)} \qquad \Delta G^{(i)} \triangleq G\left[U^{(i)}\right] - G\left[U^{(i-1)}\right].$$

7: Go to 1.

8: Accept sequence of future control inputs and terminate.

$$f^{(i)^T}G\left[U^{(i)} + \mu f^{(i)}\right] \geq \beta f^{(i)^T}G\left[U^{(i)}\right] \tag{3.167}$$

The quantity δ must be a small positive number, $\delta < 0.5$. It is natural to apply a value of the same magnitude as one D/A-converter quant since a higher precision is meaningless (selecting the δ in this way does not guarantee that this precision is obtained, but generally it will be close). β should be chosen so that $\beta \in (\delta, 1)$. A typical choice is a number close to 1, e.g., $\beta = 0.9$.

As for the Levenberg-Marquardt algorithm, implementing the Quasi-Newton algorithm in a real-time system demands certain additional constraints on the algorithm. In practice it is necessary to impose a limit on the number of times the criterion and gradient are evaluated. Although it should not occur that the loops *1-4-1* and *5-6-7-9-5* are executed more than a few times, it is nevertheless recommended to impose a limitation for safety reasons.

Table 3.7. Algorithm for selecting the step size.

0: $\mu = 1$, $I = [b_1, b_2] = [0, 1]$; $J_{b_1} = J(U^{(i)})$; $G_{b_1} = G[U^{(i)}]$.

1: Evaluate the criterion $J_{b_2} = J(U^{(i)} + \mu f^{(i)})$ and determine the gradient $G_{b_2} = G[U^{(i)} + \mu f^{(i)}]$.

2: If (3.166) and (3.167) both are satisfied go to 10.

3: If (3.166) is not satisfied go to 5.

4: $I = [\mu, 2\mu]$; $\mu = 2\mu$; $J_{b_1} = J_{b_2}$; $G_{b_1} = G_{b_2}$. Go to 1.

5: Determine β as the extremum of the second-order polynomial $P(x) = p_2 x^2 + p_1 x + p_0$ possessing the following properties: $P(b_1) = J_{b_1}$; $P'(b_1) = f^{(i)^T} G_{b_1}$, and $P(b_2) = J_{b_2}$, which is equivalent to:

$$p_2 = \frac{J_{b_2} - J_{b_1} + f^{(i)^T} G_{b_1} (b_1 - b_2)}{(b_1 - b_2)^2}$$

$$p_1 = f^{(i)^T} G_{b_1} - 2 p_2 b_1$$

$$p_0 = J_{b_1} - p_2 b_1^2 - p_1 b_1$$

That is, β is then determined by

$$\beta = -\frac{p_1}{2 p_2}$$

6: If $\min(\beta - b_1, b_2 - \beta) \geq 0.1(b_2 - b_1)$ then $\mu = \beta$. Otherwise $\mu = (b_1 + b)/2$.

7: Evaluate criterion $J(U^{(i)} + \mu f^{(i)})$ and determine the gradient $G[U^{(i)} + \mu f^{(i)}]$.

8: If (3.166) and (3.167) are both satisfied then go to 10.

9: If (3.166) is satisfied then set $I = [\mu, b_2]$; $J_{b_1} = J(U^{(i)} + f^{(i)})$; $G_{b_1} = G[U^{(i)} + f^{(i)}]$ and go to 5. Otherwise set $I = [b_1, \mu]$; $J_{b_2} = J(U^{(i)} + \mu f^{(i)})$; $G_{b_2} = G[U^{(i)} + \mu f^{(i)}]$ and go to 5.

10: Accept step and return to main algorithm.

Tuning the Controller. At this stage the basic implementation issues have been covered and the attention is now drawn towards the practical use of the predictive controller. Soeterboek (1992) contains a quite elaborate discussion on how to tune predictive controllers. This is widely supported by illustrative simulation studies to give the reader a feeling for the impact a given choice of design parameters has on the control. It is interesting to remark that as opposed to the previously mentioned design methods, the computation time required by the nonlinear predictive controller is heavily influenced by the choice of design parameters. The two horizons N_2 and N_u have a substantial impact on the time needed to compute the control inputs.

To mention a few essential guidelines, the design parameters, N_1, N_2, N_u, and ρ should be selected as suggested in the following:

Minimum prediction horizon N_1: This is always selected to the model time-delay, d. There is no reason for choosing it smaller because the $d-1$ first predictions depend upon past control inputs only and thus cannot be influenced. On the other hand it is not recommended to choose it bigger since this can lead to quite unpredictable results (Soeterboek, 1992).

Maximum prediction horizon, N_2: To ensure stabilization of systems with an unstable inverse, it should be at least as many time steps as there are past control inputs fed into the network model. Usually it is selected somewhat longer. A rule of thumb says that the prediction horizon should be selected approximately as the rise time of the system (if it is stable). However, often it is not possible to choose it this long because the optimization problem will become too demanding compared with the selected sampling period.

Control horizon, N_u: In the linear case it is recommended to let it equal or exceed the number of unstable or poorly damped poles. Indeed this information can be difficult to retrieve if only a neural network model is available and one only possesses little physical insight about the dynamics of the system. Using instantaneous linearization to produce a PZ map as the one displayed in Figure 3.22 may in this case be quite valuable. As an overall valid value, Soeterboek (1992) suggests to simply use $N_u = n$. However, in the present case where the computational burden increases tremendously as N_u is made longer, this is probably a somewhat large horizon in many cases. Unless the sampling period is very long it is thus recommended to always use the smallest viable value.

The control penalty factor, ρ: For reasons of numerical robustness, it is recommended to use $\rho > 0$ to prevent the Hessian from becoming singular due to several feasible solutions (Soeterboek, 1992). However, it is primarily used for controlling magnitude and smoothness of the control signal and in practice it can be selected from simulation studies.

3.8.2 NPC Applied to the Benchmark System

Now NPC control of the benchmark system will be studied. The objective is now to control the benchmark system by nonlinear generalized predictive control. The proposed Newton-based Levenberg-Marquardt method is applied for minimization of the criterion. To reduce the amount of computations, the algorithm is stopped after 5 iterations or when $|U^{(i+1)} - U^{(i)}| < 10^{-4}$. After a little experimenting, the following design parameters were selected:

$$N_1 = d = 1 \,, \qquad N_2 = 7 \,, \qquad N_u = 2 \,, \qquad \rho = 0.01 \,.$$

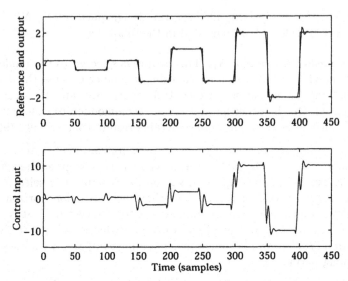

Figure 3.26. Nonlinear predictive control of the benchmark system.

The result of the simulation is displayed in Figure 3.26.

It appears that a quite rapid response to reference changes is achieved while a smooth control signal of modest magnitude is maintained. Notice also how the step changes are anticipated by the controller. This is because the control is based on future references. It is also interesting to note that the response shows a non-minimum phase behavior although the inverse of the system is stable (output shoots the "wrong" way). This is not uncommon when a system is controlled by a predictive controller (Soeterboek, 1992).

In comparison to the model-following techniques (direct inverse control, feedback linearization, pole placement, etc.) it is interesting to note how criterion-based control of nonlinear systems leads to different response shapes for step changes of different sizes. The reason for this being that the closed-loop system is not linear in this case.

3.8.3 Approximate Predictive Control (APC)

From a computational perspective, the nonlinear predictive controller has quite comprehensive demands. With the electronic off-the-shelf hardware available today, this will probably make the strategy unrealistic for control of systems with dynamics that are not reasonably slow. In addition, a number of precautions typically imply that certain *ad hoc* solutions need to be introduced for robustness reasons. This will in turn increase the complexity of the control law further. Typical *ad hoc* solutions must address:

- Has the iterative minimization algorithm converged to an acceptable accuracy?

- Do numerical problems occur?

- Is the found solution actually the global minimum of the criterion?

- How many times should one re-execute the minimization algorithm from different initial points?

Thus, applying the principle of instantaneous linearization to predictive control design will give tremendous advantages over the conventional nonlinear design examined previously. When the model is linear (and $\rho > 0$), the criterion has a unique minimum which can be found directly. Hence, using a design based on a linearized model will remedy all the problems mentioned above. Also it is straightforward to modify the design for stochastic models as well. As mentioned in the derivation of the instantaneous linearization principle, the linearized models are, however, generally only valid near the operating point. For systems with non-smooth nonlinearities there will be problems.

The Criterion. To account for the operating point dependent bias component, $\zeta(\tau)$, (see Section 3.7) as well as for regular constant disturbances , it is desirable to let the controller include integral action. To achieve this it is necessary to: 1) consider a criterion where the differenced control inputs are weighted and 2) select the predictor carefully.

Consider again the GPC criterion

$$J(t, U(t)) = \left[R(t) - \hat{Y}(t) \right]^T \left[R(t) - \hat{Y}(t) \right] + \rho \tilde{U}^T(t)\tilde{U}(t) . \qquad (3.168)$$

To simplify the derivations it is now assumed that N_1 equals the time delay, d.

The Predictor. It is assumed that an ARX model has been obtained by instantaneous linearization of a NNARX model

$$A(q^{-1})y(t) = q^{-d}B(q^{-1})u(t) + e(t) + \zeta(\tau) . \qquad (3.169)$$

Interpreting the bias term as integrated white noise, an integrated ARX model (ARIX) is achieved

$$A(q^{-1})y(t) = q^k B(q^{-1})u(t) + \frac{e(t)}{\Delta} . \qquad (3.170)$$

If the k-step ahead predictor is chosen as the minimum variance predictor, this corresponds exactly to the set-up considered in Clarke et al. (1987a). The ARIX model is examined here for simplicity and because it is considered to be the most important in practice. Clarke et al. (1987b) also derive the predictor for ARIMAX models and in Soeterboek (1992) an even more general approach is taken in order to derive predictors for systems affected by different deterministic disturbances and noise sources.

Considering the time instant $t + k$, the ARIX model equivalently reads

$$\Delta A(q^{-1})y(t + k) = q^{-d}B(q^{-1})\Delta u(t + k) + e(t + k) \,. \qquad (3.171)$$

To systematically derive a predictor, the model is now reorganized by introduction of the following Diophantine equation

$$1 = \Delta A(q^{-1})E_k(q^{-1})q^{-k}F_k(q^{-1}) \qquad (3.172)$$

$$\deg(E_k) = k - 1 \,, \qquad \deg(F_k) = n_a \,.$$

Multiplying $E_k(q^{-1})$ to both sides of (3.171) and using the identity (3.172) gives

$$y(t + k) = q^{-d}E_k(q^{-1})B(q^{-1})\Delta u(t + k) + F_k(q^{-1})y(t) + E_k(q^{-1})e(t + k) \,.$$
$$(3.173)$$

Given that the sequence of future control inputs is known, the minimum variance predictor for $y(t + k)$ is the expectation conditioned on the information gathered up to time t, I_t, (Ljung, 1999):

$$\hat{y}(t + k) \triangleq \hat{y}(t + k|t) = q^{-d}E_k(q^{-1})B(q^{-1})\Delta u(t + k) + F_k(q^{-1})y(t)$$
$$= G_k(q^{-1})\Delta u(t + k - d) + F_k(q^{-1})y(t) \,. \quad (3.174)$$

$G_k(q^{-1}) = E_k(q^{-1})B(q^{-1})$ is clearly a polynomial of order $n_b + k - 1$. By multiplying $B(q^{-1})/\Delta A(q^{-1})$ to both sides of the identity (3.172) it is obvious that the first k terms in $G_k(q^{-1})$ are the first k coefficients of the step response of $B(q^{-1})/\Delta A(q^{-1})$. Consequently the $k-1$ first terms in $G_k(q^{-1})$ must equal the $k - 1$ first terms in $G_{k-1}(q^{-1})$.

The only unknown quantities in (3.174) are now the future control inputs. In order to derive the control law it is necessary to separate these from the part of the expression containing known (=past) data

$$\hat{y}(t + k) = \begin{cases} G_k(q^{-1})\Delta u(t + k - d) + F_k(q^{-1})y(t) \,, & \text{for } 1 \le k < d \\[2mm] g_0\Delta u(t) + \left[G_d(q^{-1}) - g_0\right]\Delta u(t) + F_d(q^{-1})y(t) \,, & \text{for } k = d \\[2mm] \bar{G}(q^{-1})\Delta u(t + k - d) \\ \quad + q^{k-d}\left[G_k(q^{-1}) - \bar{G}_k(q^{-1})\right]\Delta u(t) \\ \quad + F_k(q^{-1})y(t) \,, & \text{for } d < k \le N_2 \end{cases}$$
$$(3.175)$$

where

$$\bar{G}(q^{-1}) = g_0 + g_1q^{-1} + \; \cdots \; + g_{k-d}q^{d-k} \qquad (3.176)$$

The remaining problem is now to solve all the Diophantine equations. As it turns out, this is far simpler than it initially may appear.

Since $\tilde{A}(q^{-1}) \triangleq \Delta A(q^{-1})$ is monic, the solution to

$$1 = \tilde{A}(q^{-1})E_1(q^{-1}) + q^{-1}F_1(q^{-1}) \tag{3.177}$$

is obviously

$$E_1(q^{-1}) = 1 \qquad F_1(q^{-1}) = q\left[1 - \tilde{A}(q^{-1})\right] \tag{3.178}$$

Assume now that the solution to

$$1 = \tilde{A}(q^{-1})E_k(q^{-1}) + q^{-k}F_k(q^{-1}) \tag{3.179}$$

for some k exists, and consider the equation for $k+1$

$$1 = \tilde{A}(q^{-1})E_{k+1}(q^{-1}) + q^{-(k+1)}F_{k+1}(q^{-1}) . \tag{3.180}$$

Subtracting the two gives

$$\begin{aligned}
0 = &\tilde{A}(q^{-1})\left[E_{k+1}(q^{-1}) - E_k(q^{-1})\right] \\
&+ q^{-k}\left[q^{-1}F_{k+1}(q^{-1}) - F_k(q^{-1})\right]
\end{aligned} \tag{3.181}$$

since $\deg(E_{k+1} - E_k) = \deg(E_{k+1}) = k$ it is turns out to be a good idea to define

$$E_{k+1}(q^{-1}) - E_k(q^{-1}) = \bar{E}_{k+1}(q^{-1}) + e_k^{(k+1)}q^{-k} , \tag{3.182}$$

where $e_k^{(k+1)}$ specifies the coefficient to q^{-k} in the polynomial E_{k+1}. Using this, (3.181) can be rewritten as

$$\begin{aligned}
0 = &\tilde{A}(q^{-1})\bar{E}_{k+1}(q^{-1}) \\
&+ q^{-k}\left[q^{-1}F_{k+1}(q^{-1}) - F_k(q^{-1}) + \tilde{A}(q^{-1})e_k^{(k+1)}\right]
\end{aligned} \tag{3.183}$$

By again exploiting that $\tilde{A}$ is monic, it is evident that $\bar{E}_{k+1}(q^{-1}) = 0$, leading to

$$E_{k+1}(q^{-1}) = E_k(q^{-1}) + e_k^{(k+1)}q^{-k} . \tag{3.184}$$

Consequently,

$$q^{-1}F_{k+1}(q^{-1}) - F_k(q^{-1}) + \tilde{A}(q^{-1})e_k^{(k+1)} = 0$$

or

$$F_{k+1}(q^{-1}) = q\left[F_k(q^{-1}) - \tilde{A}(q^{-1})e_k^{(k+1)}\right] . \tag{3.185}$$

Again, due to $\tilde{A}$ being monic,

$$e_k^{(k+1)} = f_0^{(k)} \, . \tag{3.186}$$

From (3.182) it also follows that

$$G_{k+1}(q^{-1}) = B(q^{-1})E_{k+1}(q^{-1}) = G_k(q^{-1}) + q^{-k}B(q^{-1})f_0^{(k)} \, . \tag{3.187}$$

As discussed above, the first k terms of G_{k+1} are thus identical to those of G_k.

According to Clarke et al. (1987a), this method is referred to as *recursion of the Diophantine equation*. The recursions are summarized in Table 3.8.

Table 3.8. Recursion of the Diophantine equation.

1: Initialize the recursions by setting $E_1(q^{-1}) = 1$ and

$$F_1(q^{-1}) = q\left[1 - \tilde{A}(q^{-1})\right]$$

Set k=1.

2: $e_k^{(k+1)} = f_0^{(k)}$ and for $i = 0, \dots, n_a$

$$f_i^{(k+1)} = f_{i+1}^{(k)} - \tilde{a}_{i+1}f_0^{(k)} \, ,$$

where $f_{n_1+1}^{(k)} = 0$.

3: if $k < N_2$, $k = k + 1$, go to *Step 2*.

Deriving the Control Law. First the predictions derived above are expressed in the following vector notation

$$\hat{Y} = \Gamma\tilde{U} + \Phi \tag{3.188}$$

where

$$\hat{Y} = [\hat{y}(t+d) \ \hat{y}(t+d+1) \ \dots \ \hat{y}(t+N_2)]^T$$
$$\tilde{U} = [\Delta u(t) \ \Delta u(t+1) \ \dots \ \Delta u(t+N_u-1)]^T$$
$$\Phi = [\varphi(t+d) \ \varphi(t+d+1) \ \dots \ \varphi(t+N_2)]^T$$

with

$$\varphi(t+k) = q^{k-d}\left[G_k(q^{-1}) - g_0 - g_1q^{-1} \ \dots \ - g_{k-d}q^{d-k}\right]\Delta u(t)$$
$$+F_k(q^{-1})y(t) \, . \tag{3.189}$$

Γ is a matrix of dimension $(N_2 - d + 1) \times N_u$:

$$
\Gamma = \begin{bmatrix}
g_0 & 0 & \cdots & 0 \\
g_1 & g_0 & & 0 \\
 & g_1 & & \vdots \\
\vdots & \vdots & & g_0 \\
g_{N_2-d} & g_{N_2-d-1} & \cdots & g_{N_2-d-N_u+1}
\end{bmatrix} . \tag{3.190}
$$

After insertion of (3.188) the criterion reads

$$
J(t, U(t)) = \left[R - \Gamma \tilde{U} - \varPhi \right]^T \left[R - \Gamma \tilde{U} - \varPhi \right] + \rho \tilde{U}^T \tilde{U} . \tag{3.191}
$$

The sequence of future control inputs is determined by setting the derivative of the criterion equal to zero

$$
\frac{\partial J(t, U(t))}{\partial \tilde{U}} = 2\Gamma^T \Gamma \tilde{U} - 2\Gamma^T (R - \varPhi) + 2\rho \tilde{U} = 0
$$

or

$$
\tilde{U} = \left[\Gamma^T \Gamma \rho I_{N_u} \right]^{-1} \Gamma^T (R - \varPhi) \tag{3.192}
$$

Only the vector $\varPhi$ will depend on the present observation of the output. Most of the computations can therefore be executed in the previous sample. Since the matrix to be inverted is symmetric and positive definite, the inversion can be carried out using Cholesky factorization. Naturally, it is irrelevant to calculate the entire vector $\tilde{U}$ since only the first entry, $\Delta u(t)$, is needed. The only calculations needed between A/D and D/A-converter calls will therefore be to finish the calculation of $\varPhi$, and a scalar product between two vectors.

Tuning the Controller. The tuning rules outlined previously are still valid, but the computational demands associated with selecting long prediction and control horizons are significantly less pronounced in the present case. However, for systems where the nonlinearities are not reasonably smooth, it can be argued that selecting a long prediction horizons is absurd. A remote future prediction will be completely unreliable and using it for calculating the present control action intuitively does not make sense.

3.8.4 APC applied to the Benchmark System

The example from before is now repeated for the APC. The design parameters were selected almost as in the previous example:

$$
N_1 = d = 1 , \qquad N_2 = 7 , \qquad N_u = 2 , \qquad \rho = 0.03 .
$$

The result of the simulation is shown in Figure 3.27.

The nonlinearity is quite smooth for this simple system, and the performance thus matches that of the NPC. However, much fewer computations were

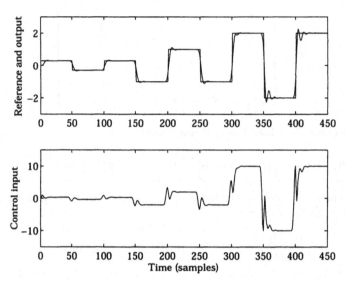

Figure 3.27. Approximate generalized predictive control of the benchmark system.

required to compute the control inputs. One cannot expect the responses to be exactly similar because the predictions have been calculated in completely different ways.

Notice that the weight on differenced control inputs, ρ, was selected higher in this example than in the previous. By demanding an increased smoothness of the control input, abrupt changes of the operating point are avoided.

3.8.5 Extensions to the Predictive Controller

Numerous variations on the GPC criterion exist, including simplifications as well as augmentations. In Soeterboek (1992) an attempt has been made to unify everything into the very general *Unified Predictive Controller* or *UPC*. The UPC offers a manifold of degrees of freedom for tuning the controller and also accommodates compensation for disturbances with known characteristics. Still, the price paid in terms of a more complex implementation is not overwhelming. The unified criterion takes the form

$$J(t, U(t)) = \sum_{k=N_1}^{N_2} \left[P(1)r(t+k) - P(q^{-1})\hat{y}(t+k) \right]^2$$
$$+ \rho \sum_{k=1}^{N_u} \left[\frac{Q_n(q^{-1})}{Q_d(q^{-1})} u(t+k-1) \right]^2 , \qquad (3.193)$$

which is minimized with respect to the N_u future control inputs and subject to the constraint

$$\phi(q^{-1})P(q^{-1})u(t+k) = 0 , \qquad N_u < k \le N_2 - d . \qquad (3.194)$$

The polynomial $P(q^{-1})$ here plays a role which relates to model-following, and it is used for tuning the servo property of the controller. It can also help reducing possible overshoot (Clarke et al., 1987b). $\phi(q^{-1})$ is a polynomial in the delay operator that when applied to the disturbance signal gives zero:

$$\phi(q^{-1})v(t) = 0 . \qquad (3.195)$$

For constant disturbances $\phi(q^{-1}) = \Delta$ as used throughout Section 3.8. The choice of $\phi(q^{-1})$ was discussed in Section 3.7.3. $Q_n(q^{-1})$ and $Q_d(q^{-1})$ are incorporated primarily to compensate for disturbances with known characteristics. Q_n should therefore include $\phi(q^{-1})P(q^{-1})$ as a factor.

One of the most appealing features of predictive control is probably the natural fashion in which constraints on the controller performance are incorporated into the criterion. Such constraints on the design could for example include saturation and rate limits in the actuator or limitations on the accepted behavior of the output. In Soeterboek (1992) the presentation focuses on actuator limitations. Solving this kind of constrained optimization problem for nonlinear models will make an already immense task almost staggering (it has, however, been done in Haley et al. (1999)). In the context of approximate predictive control it seems significantly less overwhelming.

Recently some results on stability for predictive controllers based on nonlinear models have appeared. See for example Mayne and Michalaska (1990) and Eaton et al. (1994).

3.8.6 Section Summary

This section presented two different approaches to predictive control of nonlinear systems modelled by neural networks. In the *Nonlinear Predictive Controller (NPC)* the predictions were calculated by recursion of the one-step ahead neural network predictor. This made calculation of the control input very hard as it had to be determined with an iterative minimization algorithm deployed at each sample. Two minimization algorithms were proposed that were considered particularly appropriate for the problem. The second of the proposed controllers was called the *Approximate Predictive Controller (APC)*. It was based on linearized models "extracted" from the neural network model by using the instantaneous linearization principle. This made calculation of control inputs much simpler as the minimum of the controller criterion was unique and easy to find.

The general recommendation was that the APC strategy should be preferred over the NPC. The implementation was much simpler and computationally it was considerably less demanding. The most important objection against the APC was that problems might occur if the approximate models obtained with instantaneous linearization are accurate only in the absolute proximity of the current operating point. Also the linearization technique may suffer from a sensitivity to noise.

In case the APC is not successful for the application under consideration, it may be used for initializing the iterative minimization algorithm used by the NPC. By doing so, one will achieve a performance that is equally good or exceeds that the APC.

The most important characteristics of the two predictive control strategies are summarized below:

Advantages:

- Can be tuned by quite intuitive means.
- The criterion is flexible which enables control of a wide class of systems.
- Good for controlling systems with time delay.
- Can stabilize unstable systems or systems for which the inverse in unstable.
- Compensation for disturbances with known characteristics.
- It is possible to take into account constraints on, e.g., input and output.

Disadvantages, NPC:

- The criterion may have several local minima.
- Computationally demanding.
- *Ad hoc* fixes for handling stopping, numerical problems, local minima, etc., might conflict with the real-time demands.

Disadvantages, APC:

- Relies on linearized models, which may have a limited validity in certain regimes of the operating range.

3.9 Recapitulation of Control Design Methods

A number of different approaches to the design of controllers for unknown nonlinear systems have now been presented. Before beginning to consider the advanced applications it is thus appropriate to recapitulate the overall issues of Chapter 3. The purpose is to outline a set of rules of thumb for selecting the most suitable control design for a specific application.

General advice

- Keep the model of the system as simple as possible. This may seem in contradiction with the validation and model structure selection techniques discussed in Chapter 2; the model with the lowest generalization error need not be the best model for control design. However, a smooth low-order model will frequently result in a superior closed-loop performance.

- A one-shot strategy where an experiment is conducted, a model identified and a satisfactory controller is designed is probably somewhat optimistic for many applications. One must rely on an iterative design methodology where the experiment is repeated in closed-loop with a preliminary designed controller.

- It is emphasized that a carefully prepared and well performed experiment for collecting data is of vital importance. It is a necessity for being able to identify a model that is sufficiently reliable to be used as a foundation for the control system design.

- Do not treat a system as multi-input, multi-output unless the cross-couplings are non-negligible.

Categorizing a system

The special characteristics of the system to be controlled must be clarified prior to selection of a design method. Instantaneous linearization obviously lends itself to assisting such an analysis. It can be carried out on the training and validation data sets subsequent to the identification procedure.

- Using instantaneous linearization it is easy to review the smoothness of the nonlinearities in the system. Depicting the coefficients of the extracted linear models, the poles, or the damping factor(s) and natural frequency(-ies) will often provide a valuable understanding of the character of the nonlinearities.

- By extracting linear models with instantaneous linearization and calculating the zeros, important information is revealed. The location of the zeros is important in relation to the feasibility and practical applicability of several of the presented controller designs.

- The linearized models can also help in verifying that the selection of the sampling frequency is sensible. When considering the coefficients of the linearized models, they may appear almost constant although the nonlinearities are severe. This is a numerical property of fast sampling as the poles and some of the zeros will converge to $z = 1$ when the sampling frequency is increased.

- It is a good idea to include in the validation phase an examination of the identified model's ability to perform multistep ahead predictions. Either by the recursion technique used in Section 3.8.1 or by the minimum-variance technique based on instantaneous linearization discussed in Section 3.8.3. This can clarify if the model is suitable for predictive control.

Rules of thumb for selecting a suitable design

Below an attempt is made to outline some guidelines for selecting a control design that is appropriate for a given application.

Smooth nonlinearities: If the nonlinearities are reasonably smooth, the primary choice is unconditionally a design method based on instantaneous linearization. Working with linear models gives a freedom in the design phase that none of the other strategies can match. Although the designs based on linearization are structurally equivalent to a self-tuning regulator and thus will require more computations than a pure neural network controller, the amount of computations are typically not overwhelming. The approximate predictive controller is a very flexible design allowing control of a large class of systems. It is fairly easy to tune and consequently it is recommended as a first choice. If one has very strict demands to the behavior of the closed-loop system, the use of an approximate pole placement design is recommended.

Nonsmooth nonlinearities: If the nonlinearities are hard, or at least so that models extracted by instantaneous linearization change dramatically with small variations in the operating point, a true nonlinear design is normally required. If the sampling frequency permits it, the nonlinear predictive controller (NPC) is normally preferred. If not, direct inverse control, internal model control, optimal control, or feedback linearization are all possible alternatives.

Known reference trajectory: Servo problems for which the same reference trajectory is constantly repeated occur in many robotics applications. The specialized training of inverse and optimal controllers is useful for optimizing a controller specifically for such a pre-specified trajectory. An alternative, which perhaps possesses the biggest industrial potential, is the neural-network-based feedforward. This is added to an existing feedback control system in order to enhance the performance. It has the advantage that the sequence of feedforward control inputs can be calculated in advance, thus making implementation extremely simple. For example it makes it possible to implement neural network control in very simple digital systems. If the reference trajectory is modified, the network is re-evaluated and the new feedforward sequence is incorporated into the controller program. An important advantage of the neural network feed-

forward is that it does not interfere with the stability of the closed-loop
system.

Regulation: For regulation problems there are different suitable options. If
no major disturbances are acting on the system, a steady-state type of
feedforward control can enhance the response to set-point changes while
a robust stabilization of the system is maintained with a PID controller.
If unmeasured disturbances with known characteristics are acting on the
system, an approximate pole placement (or predictive controller) is recommended. For these controllers it is relatively easy to incorporate compensation for disturbances in the design.

Noise acting on system and/or measurements: The approximate MV1 and
GPC controllers were designed for controlling stochastic systems under
the assumption that the noise entered the system in a particular fashion.
However, since instantaneous linearization is based on differentiation of a
network model, one must expect a certain sensitivity to noise. It may be
a good idea to either low-pass filter the signals before feeding them into
the neural network or, alternatively, to low-pass filter the coefficients of
the extracted linear models. If it is possible to stabilize the system with a
PI(D)-controller and the conditions for controlling the system by an inverse model are satisfied, using a neural network feedforward to optimize
the control system may turn out to be the most robust solution.

Fast dynamics: If the system requires a sampling frequency which is high
compared with the computing power of the real-time hardware in which
the controller is implemented, it will prohibit application of some of the
more complex controller designs. The simplest controllers are those directly realized by a neural network; e.g., direct inverse control, optimal
control, and feedback linearization. Unfortunately, some of the algorithms
used for training these controllers must be executed on-line and are thus
quite demanding, e.g., specialized training. A remedy is to use a faster
computer in the training phase if this is available. The methods based
on instantaneous linearization are typically more demanding than pure
neural network controllers since they require a redesign of the controller
in each sample. However, most of the computations can be performed
after D/A-conversion so that the amount of computations between A/D
and D/A-conversions is minimal.

Slow dynamics: If the sampling frequency is adequately low to allow for an
implementation of all the concepts presented in this chapter, the neural
network controllers will usually not be preferred. The most demanding
strategies, NPC and APC, are probably also the ones most likely to give
a robust stabilization and a high performance.

Multi-input, multi-output systems (MIMO): Although only SISO systems have
been examined in this chapter, some of the proposed designs are easily

extended to cover the MIMO case as well. For other designs it is more involved. Tuning tends to become a particularly complicated matter for the MIMO systems. In general a criterion-based design is the best choice. Since extention of the nonlinear predictive controller to the MIMO case will be almost overwhelming, a criterion-based design relying on instantaneous linearization is perhaps the most realistic solution. For example, one can use an approximate predictive controller or an approximate linear quadratic regulator.

Time delay: If the system includes a time delay exceeding one sampling period, the recommended choice of design is one of the predictive control strategies.

Unstable: If the system is unstable it cannot be controlled with an internal model controller. Otherwise all the design methods covered in the chapter are potential candidates. If on-line training is employed, one must anticipate problems if the neural network is not initialized to provide a stabilizing control.

Unstable inverse: The methods developed to cancel the nonlinearities, e.g., feedback linearization and control involving inverse models, cannot be used on systems with an unstable or poorly damped inverse. Neither can an approximate pole placement with full zero cancellation. This leaves predictive control and approximate pole placement without zero cancellation as the most appropriate designs for this particular class of systems.

4. Case Studies

As a conclusion, this chapter will show four examples of the use of neural networks for modelling and control. Each example covers selected issues that the user of neural network techniques will often face. It is by no means claimed that the "optimal" solution is found in any of the applications; they serve primarily as proof of concepts. That is, neural networks are simple means for solving challenging nonlinear problems.

The examples have been selected so that the reader should easily be able to repeat them or use them for trying some of the other possible solutions described in this book. In all the examples, the MATLAB®-based neural network software mentioned in the Preface has been used.

The first two case studies are examples of system identification while the subsequent two are control problems.

Sunspot activity (Section 4.1). This example is a classic benchmark test used in the time series analysis community. The data set contains the averaged annual sunspot activity over more than two centuries. The data set is used to demonstrate how pruning and regularization can improve the generalization ability of neural network models.

Hydraulic actuator for a crane (Section 4.2). This example features a hydraulic actuator for a crane. It will be demonstrated how a neural network model can significantly outperform a linear model despite the fact that one might at first be led to believe that the linear model is quite good. It will be illustrated that when the system is sampled fast compared with the dynamics of the system, one should be more careful in the validation.

Pneumatic position servomechanism (Section 4.3). To demonstrate the alleged advantages of the two predictive control designs proposed in Chapter 3 they will be evaluated in an advanced simulation study of a pneumatic position servomechanism. The servomechanism exhibits a quite severe nonlinear behavior and thus it represents a reasonable challenge for the neural-network-based controllers.

Level in a water tank (Section 4.4). The final example will show the direct inverse control scheme applied to the control of a water level in a conic shaped tank. The water tank has been given a conic shape to make the nonlinearities particularly pronounced.

4.1 The Sunspot Benchmark

The first case study is an example of nonlinear time series modelling with neural networks. The data set in question consists of the average annual sunspot activity collected over almost three centuries. The data set has for long been one of the most common benchmarks in the time series community, and different approaches to modelling of the underlying process can be found in several publications, e.g., Priestley (1988), Weigend et al. (1990), Tong (1990), Svarer et al. (1993). In the present case it will be shown how the Optimal Brain Surgeon (OBS) algorithm can be used for automatic selection of a (nearly) optimal model structure. The algorithm can assist the user in selecting the internal network architecture as well as the regressors used as inputs to the network.

The complete data record is shown in Figure 4.1. It covers the period 1700–1979. Due to common practice, the period 1700–1920 is used for training while the periods 1921–1955 and 1956–1979 represent two separate test sets. The data have been scaled so that the activity is between 0 and 1. Another

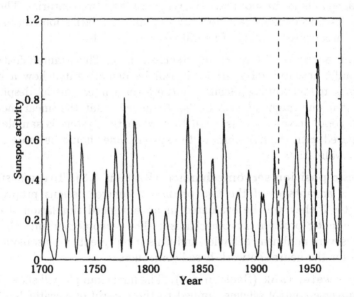

Figure 4.1. Scaled record of sunspot activity for the period 1700–1979. The training set is the period 1700–1920, test set 1 is the period 1921–1955, and test set 2 is the period 1956–1979.

common practice is to model the time series as an AR-model; i.e., it is assumed that the activity in the present year is best predicted by a function of the activity a number of years back.

4.1.1 Modelling with a Fully Connected Network

A neural network AR-model will now be trained. The model structure is deliberately selected quite large as pruning will later be used for reducing the structure to a more appropriate size. The sunspot activity appears to have a period of roughly 11 years, but the regressors are initially selected so that the current year is predicted from the past 12 years. The OBS algorithm should then be able to show that it is sufficient to input only the past 11 years. The initial, fully connected network has 8 hidden units, corresponding to 113 weights. See Figure 4.2.

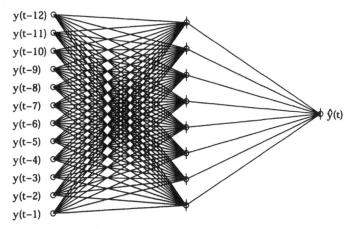

Figure 4.2. The initial, fully connected model structure. The sunspot activity is described by an NNAR(12) model. The network has 8 hidden *tanh* units and one linear output unit.

In order to assess the accuracy of the neural network models, their performance will be evaluated on the three data sets according to the scaled error:

$$E_{set} = \frac{1}{N\sigma_{total}^2} \sum_{k=1}^{N} [y(t) - \hat{y}(t)]^2 \; , \tag{4.1}$$

where σ_{total}^2 is the variance of the complete data set. This is to enable comparison with treatments of the time series found elsewhere.

First the fully connected network will be trained. The data set is scaled to zero mean and unit variance prior to the training. After the training, the weights in the network will be rescaled to allow the model to be applied to unscaled data. The network is trained with regularization by using a small weight decay (much smaller than the optimal value). The small weight decay makes it possible to invert the Hessian of the criterion. This is necessary for being able to employ the OBS algorithm. The weight decay also enforces a certain smoothness in the criterion, thus making the OBS more reliable. A typical network trained with the weight decay $\alpha = 10^{-4}$ has the scaled errors listed in Table 4.1.

Table 4.1. Scaled errors for fully connected network.

Train (1700–20)	Test (1921–55)	Test (1956–79)
0.016	0.235	0.931

4.1.2 Pruning of the Network Architecture

Due to the large amount of adjustable parameters ($p = 113$) compared to the training set size ($N = 221 - 12 = 209$) some overfitting must be expected. In order to improve the generalization, the network will now be pruned according to the OBS algorithm. Initially the input-to-hidden layer weights are pruned one at a time. When a unit is connected to only a single input, the saliency for the entire unit is calculated. When the unit saliency is smaller than any other saliency, the entire unit is removed. For the best result, the network is retrained each time a weight is eliminated. The pruning session is shown in Figure 4.3.

As one would anticipate, the error on the training set is getting larger as the network architecture is pruned. The two test errors, on the other hand, are initially decreased but at some point they start growing again. It is seen that the decrease in the test errors is quite substantial. Thus, a much better performance is apparently gained by the pruning.

It is a less encouraging result that the FPE estimate is surprisingly different from the test errors. It should be emphasized that test error and FPE do not estimate the same quantity; the test error estimates the generalization error, wheras FPE averages over all data sets of size N. Nevertheless, it appears to be a general result that the FPE underestimates the average generalization error. This is mainly attributed to the fact that the estimate has been derived based on overly crude approximations.

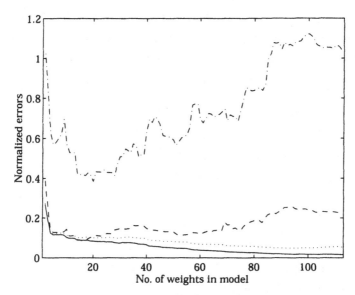

Figure 4.3. The pruning session. The curves are read from right to left. The errors to the right are associated with the initial network (113 weights). The figure shows training error (solid), first test error (dashed), second test error (dash-dotted), and final prediction error estimate (dotted).

Notice that the two test errors are surprisingly far apart. As the data sets are rather small, the estimates are of course sensitive to noise, but it is also believed that the time series is in fact nonstationary. This can also explain why the model generalizes better to the first test set than to the second.

As the "optimal" model structure, the network with 17 weights will be picked. The first test set reaches its minimum here, and the second test set is fairly close to its minimum. When the network is trained without weight decay and evaluated on the three data sets, the results listed in Table 4.2 are obtained.

Table 4.2. Scaled errors for pruned network.

Train (1700–20)	Test (1921–55)	Test (1956–79)
0.087	0.084	0.414

These results are clearly better than for the initial network. In Svarer et al. (1993) a comparison is made between different analyses of the same time series. The results obtained above are fairly close to the results listed in this reference, although not quite as good. Following the approach in Svarer et al.

(1993), a better result can probably be obtained with a better choice of weight decay, but this will also be much more time-consuming.

The architecture of the final neural network is plotted in Figure 4.4. The

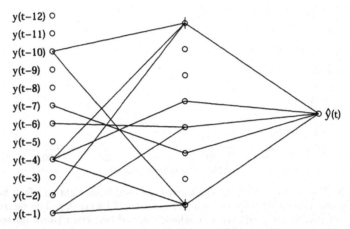

Figure 4.4. The final model structure. Only 5 hidden units are active and only 17 weights are left in the network. A vertical line through a unit symbolizes a bias (which is also counted as weight).

figure shows that the oldest input, $y(t-12)$, has been removed as expected. More surprisingly, the input $y(t-11)$ has been removed as well. As the reader quickly will experience when working with the pruning algorithms, the result of the pruning session can vary a great deal. In some cases $y(t-11)$ survives in the optimal structure and in other cases even $y(t-12)$ remains. The conclusion is that one should always run multiple pruning sessions, each one started with a different set of network weights.

It should be noticed that the results of this case study have been achieved by making "active" use of the test sets. In Tong (1990) a distinction is made between this type of result, and the so-called *genuine predictions*. In the latter case, the final model is found using only the data in the training set. The test sets are strictly used for validation. This distinction is important if a reliable estimate of the generalization error is to be obtained.

4.1.3 Section Summary

In this section the pruning algorithm *Optimal Brain Surgeon* (OBS) was employed to select an appropriate model structure for the sunspot benchmark problem. Some of the conclusions drawn in the example were that:

- It is possible to obtain good results for the sunspot time series with model structures based on neural networks.

- Pruning can optimize the model structure automatically. If the model is heavily overparametrized, the OBS algorithm can lead to a substantial improvement in the generalization ability.

- The final prediction error estimate (FPE) appears to underestimate the average generalization error and in the present case it was not appropriate to use for model structure selection.

- The result of the pruning session depends heavily on the initial network. For the best results it is recommended to run multiple sessions.

4.2 Modelling of a Hydraulic Actuator

The next example is about modelling of a hydraulic actuator that is used for controlling the position of a crane arm. The crane, which is shown in Figure 4.5, has four actuators: boom, arm, telescopic extension, and rotation of the whole crane. The actuator considered in this example is the one used for controlling the boom (lower) joint of the arm. The actuator is controlled by

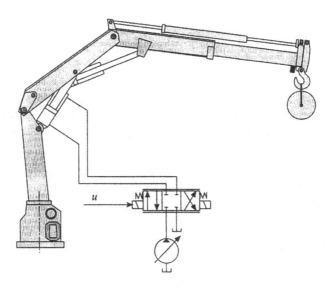

Figure 4.5. The Hydraulic Crane. (By courtesy of Svante Gunnarsson, Linköping University.)

changing the opening of the valve through which oil flows into the actuator. The amount of oil determines the pressure and in turn the actuator position. Measured values of the valve opening (input) and the oil pressure (output) are shown in Figure 4.6. Observe that the actuator exhibits a quite oscillatory response to changes in the valve opening. This behavior is due to mechanical resonances in the crane arm. A more detailed description of the set-up along

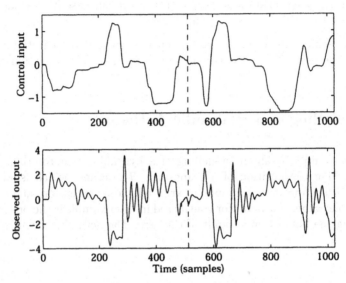

Figure 4.6. The data set; valve opening (top) and oil pressure (bottom). The first 512 samples are used for training and the remaining 512 samples for validation and model comparison.

with a physical modelling and conventional control design can be found in Gunnarsson and Krus (1997). A system identification of the actuator, using both linear and nonlinear model structures, has also been carried out in Sjöberg et al. (1995).

As for the first example, the data set is given. It is not possible to conduct additional experiments to obtain more data or to provoke a particular behavior in the system.

4.2.1 Estimation of a Linear Model

The data set is split up into a training and a test set of equal size; the first 512 samples are used for training, and the remaining 512 samples for validation and model comparison. As a first step, a linear model will be estimated. This is always a good idea because the linear model is useful as a reference against which the performance of more complicated model structures can be compared. The model structure is selected as an ARX(3,2,1) since this was recommended in Sjöberg et al. (1995). ARX(3,2,1) means that the prediction of the output is a function of the three past output observations and the two past control inputs;

$$\hat{y}(t|\theta) = -a_1 y(t-1) - a_2 y(t-2) - a_3 y(t-3) + b_1 u(t-1) + b_2 u(t-2) . \quad (4.2)$$

After removing the mean from the data, an ARX(3,2,1) model is estimated. The ability of the estimated model to predict is shown in the top panel of Figure 4.7, where it has been evaluated on test data.

It is difficult to distinguish the predictions from actual observations, which might initially lead one into believing that the model is very accurate. However, as mentioned in Chapter 2 one must be careful not to rely too heavily on a visual inspection of the one-step ahead predictions. If the system is sampled fast relative to its dynamical properties, the one-step ahead predictions will most likely appear accurate. In such cases it is a good idea to consider more distant predictions. In the lower panel of Figure 4.7 the observed output is compared with a pure simulation of the system, corresponding to an infinite prediction horizon.

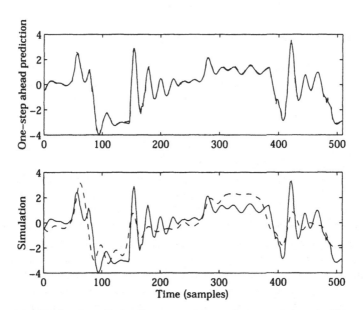

Figure 4.7. Evaluation of the linear model on the test data set. Upper panel: Observed output (solid) compared with one-step ahead predictions (dashed). Lower panel: Observed output (solid) compared with simulated output (dashed). RMS error in the latter case is 0.94.

The simulation shows that the model has captured something about the dynamical behavior of the system, but it is definitely not perfect. It appears to be particularly difficult to accurately model the oscillatory behavior of the system. In the following section an NNARX model will therefore be attempted instead.

4.2.2 Neural Network Modelling of the Actuator

Assuming that the appropriate regressors have been selected, an NNARX(3,2,1) model will now be estimated in the same way as before. The model structure is shown in Figure 4.8.

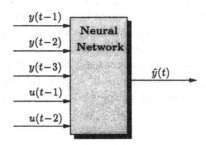

Figure 4.8. NNARX(3,2,1) model structure.

First it is necessary to get an idea about what constitutes an appropriate network architecture in the present case. Figure 4.9 shows the result of 10 different networks each trained 5 times on the same data set. The data set is scaled to zero mean and unit variance before training. Afterwards, the network weights are rescaled appropriately. The 10 network architectures are obtained by increasing the number of hidden units from 1 to 10. The general picture is as expected: the training error is decreased by making the architecture bigger; the test error is initially decreased but at some point it starts increasing again. Notice also that there is much bigger spread of the test errors for large networks. This is because more local minima are introduced when the architecture is made more flexible. It is emphasized that when this kind of model structure comparison is made, it is *absolutely vital* that the training is continued until the weights are extremely near the minimum. If the training is stopped too early, the overfitting will be less pronounced for the large model structures, thus prohibiting a fair comparison of model structures. Up to 700 iterations of the Levenberg-Marquardt algorithm were run in the present case.

It appears that a network architecture with four hidden units is the best when evaluated in this way. Following the guidelines from Section 2.7, a slightly

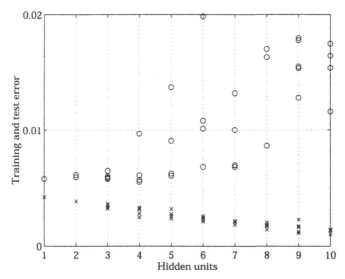

Figure 4.9. Evaluation of 50 different (fully connected) NNARX models. The network architecture is increased from having 1 hidden unit to having 10. For each architecture 5 models are trained by using different initializations of the weights. 'x' denotes the training error and 'o' denotes the test error (*error* here means the value of the criterion, i.e., one half the mean square error). Not all test errors are shown in the figure as some are outside the bounds of the y-axis. For small networks the errors tend to collide as small architectures are not subject to the problem with muliple minima.

larger network should be chosen and then one should employ either pruning or regularization to deal with the excessive use of weights. Since pruning was used in the previous example, regularization will be considered here. An experiment will now be made where a network model with eight hidden units initially is trained using the weight decay parameter, $\alpha = 0.3$. The weight decay is slowly decreased and for each value a new network is trained and evaluated. The result of the experiment is shown in Figure 4.10. It is seen that a small weight decay improves generalization, but as the weight decay is increased beyond its optimal value, the generalization deteriorates again. The error between observed and simulated output has also been included in Figure 4.10. It is seen that this error reaches its minimum almost simultaneously with the test error. However, the simulation error increases much faster with the weight decay.

It is interesting to compare the values of training and test error to those obtained when searching for a model structure (Figure 4.9). The test error, which is an estimate of the generalization error, can apparently be made smaller with regularization than can be obtained with a fully connected network architecture. This is the reason for suggesting that one initially finds the

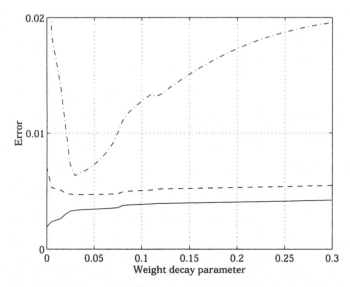

Figure 4.10. Evaluation of 61 different models obtained by varying the weight decay parameter from 0.3 to 0. Training error (solid), test error (dashed), and simulation on test set (dot-dashed). The simulation error has been divided by 50 to enable comparison.

optimal fully connected model structure, add a few extra hidden units, and then use regularization or pruning to reduce the flexibility, thereby improving generalization.

Some might argue that it is abuse of the test data when they are used actively in the modelling, which is the case when the weight decay parameter is determined as above. To get a more honest picture of the model's ability to generalize, the data used for validation should be independent of the actual training. In principle, one should have a third data set that is used strictly for validation and not for selecting the weight decay parameters. For examples like the present, where the amount of data is somewhat limited, it is difficult to spare any data for a third data set.

In Figure 4.11 a simulation is shown for a network trained with the weight decay $\alpha = 0.03$. A comparison with Figure 4.7 leaves no doubt: the accuracy of the neural network model is clearly superior to that of the linear model. In fact, the RMS error is reduced by a factor 2.5. Notice in particular how the model better describes the oscillatory behavior of the system.

If one is specifically interested in a simulation model, it would, in principle, be better to consider NNOE models as these are trained to provide an optimal simulator. However, in the present case it is surprisingly difficult to obtain useful results with NNOE models. The fact that the system has poorly damped dynamics leads to many problems when attempting to model

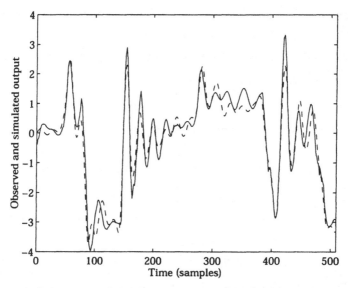

Figure 4.11. Comparison of the observed output (solid) and the simulated output (dashed). The output is simulated by a NNARX(3,2,1) model with 8 hidden units trained with regularization. The weight decay parameter was set to $\alpha = 0.03$. The RMS error between observed and simulated output is 0.38.

the system with a recurrent NNOE model. Probably it would be necessary to acquire a larger and more informative data set to be able to estimate a useful NNOE model.

4.2.3 Section Summary

The present example dealt with system identification of a hydraulic actuator for a crane. A linear model was estimated first and subsequently it was shown that a significantly better model could be achieved with the use of neural networks. The important lessons learned in the example are briefly summarized below.

- It is important not to rely too heavily on a visual inspection of the one-step ahead predictions. The predictions will generally be close to the observed outputs when the sampling frequency is high relative to the dynamics of the system. Instead one should inspect the k-step ahead predictions ($k > 1$) or make a pure simulation of the model.

- As the model structure is made larger, more local minima are introduced. When stuck in some of these minima, the generalization ability might be extremely poor.

- Regularization can effectively reduce the generalization error of the network model.

- It is better to train a network which is "too large" with regularization than to use the best of the fully connected networks.

4.3 Pneumatic Servomechanism

The next case study deals with position control of a pneumatic servomechanism. The two predictive controllers *NPC* and *APC*, which where presented in Section 3.7, will be applied to the system. It will also be explored how information about the physical system can be obtained through instantaneous linearization.

The simulation model for the pneumatic servomechanism, which will be used in the example, is a model of a laboratory set-up described in Sørensen et al. (1994). The servomechanism consists of a linear compressed air cylinder lifting an inertial weight, and the objective is to control the position of the piston using a system of servo valves. A diagram of the servomechanism is depicted in Figure 4.12. Parameter values are listed in Table 4.3.

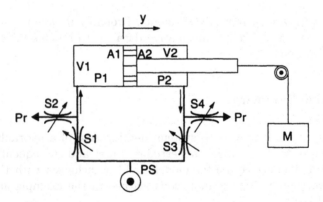

Figure 4.12. The pneumatic servomechanism.

Compared to the real set-up, certain simplifications are made in the simulation model. The most important ones are the following:

- In reality the cylinder is fed from a system of four valves that each consists of a parallel assembly of nine identical solenoid on-off valves. However, here it will be assumed that the servo valves open proportional to their control input.

Table 4.3. Parameter values in the pneumatic servomechanism model.

Mass:	$M = 20\,\text{kg}$
Area:	$A1 = 19.63\,\text{cm}^2$
Area:	$A2 = 16.49\,\text{cm}^2$
Cylinder volume:	$V1 = 0.4908\,\text{liter}$
Cylinder volume:	$V2 = 0.4123\,\text{liter}$
Static pressure level:	$P1 = 4.00\,\text{bar}$
Static pressure level:	$P2 = 5.76\,\text{bar}$
Air temperature:	$T = 293\,\text{K}$
Supply pressure:	$Ps = 6\,\text{bar}$
Return pressure:	$Pr = 1\,\text{bar}$
Valve coefficient:	$K_v = 17.142\,\text{liter/s/bar}$
Viscous friction:	$B = 30\,\text{N/m/s}$
Control input:	$S1 = S4 = U$ for $U \geq 0$
	$S2 = S3 = U$ for $U < 0$

- The valves are operated so that $S1 = S4 = U$ for $U \geq 0$ and $S2 = S3 = U$ for $U < 0$. In this way, the servomechanism can be treated as a SISO system with the common control input U.

- The stiction and Coulomb friction in the piston bearings are not modelled.

The servo valve opening characteristics are approximately linear. Since the friction is neglected, the main nonlinear behavior is therefore due to the cylinder itself. The cylinder chamber's compression dynamics are position dependent and the servo valve flow characteristic is nonlinear.

It is possible to show that a linearization of the nonlinear model around the stationary point where $y = 0$ leads to to a third-order model on the form:

$$H(s) = \frac{k}{s(s^2 + 2\zeta\omega s + \omega^2)} \tag{4.3}$$

where the natural frequency is $\omega \simeq 22\,\text{rad/s}$, and the damping factor is $\zeta = 0.034$. Apart from the integration, which obviously is present since it is a position servo, the system also has a poorly damped complex pole pair in the considered operating point.

The sampling frequency is set to $f_s = 1/T_s = 10\,\text{Hz}$.

4.3.1 Identification of the Pneumatic Servomechanism

The integration and the lightly damped pole pair makes it difficult to generate a set of data that can be used for inferring a neural network model of the system. Clearly it is necessary to operate the system in closed-loop when conducting the experiment. A manually tuned PI-controller is therefore used for stabilization of the system during the experiment. The design parameters are set to $K = 15$ and $\tau_I = 5\,\text{s}$.

To make sure that the entire operating range will be present in the training data, a high-frequency signal is applied in some periods of the experiment while in other periods the system is allowed to assume stationarity. It is also attempted to have the entire range of possible positions from -0.245 meters to $+0.245$ meters present in the data set. Depending on ones patience, as much data as desired can be collected. In this study, 3000 samples, corresponding to 300 seconds, are considered sufficient for training. In a similar fashion, a data set is produced for validation purposes. The complete training set is shown in Figure 4.13.

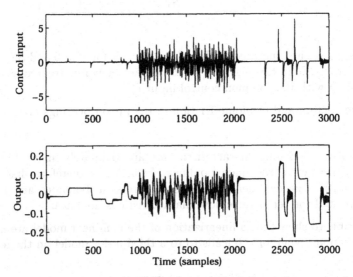

Figure 4.13. The training data. Upper panel: control signal. Lower panel: position in meters.

Since no noise or external disturbances are acting on the system, and the training set is large, it is fairly easy to identify a neural network model. Utilizing the knowledge that the system is of order four, a NNARX(4,4,1) model structure is selected. By gradually increasing the number of hidden units while evaluating the test error for the trained networks it is found that the minimum test error is achieved with 12 hidden units. 12 hidden units corresponds to 121 weights, which is a relatively small number compared to the size of the training set. Thus, it is considered unnecessary to use regularization or pruning to improve generalization.

4.3.2 Nonlinear Predictive Control of the Servo

First the "true" nonlinear predictive controller (NPC) will be considered. The design parameters in the controller are set as follows: there is no time delay in the system except for the usual delay of one sampling period. The minimum costing horizon is therefore set to one: $N_1 = 1$. The prediction horizon (N_2) is set to 10 since this value in Clarke et al. (1987a) is recommended as a good default choice for a large class of practical systems. The only design parameters left to tune are then the control horizon N_u and the penalty factor ρ. These are adjusted to achieve a response that is reasonably fast and has little or no overshoot while the control signal is kept smooth. In addition, the control inputs should be limited to the range from -5 to $+5$ since the training set contained data from this range only. The position reference is a series of steps of increasing amplitude, the final amplitude being close to the full extention of the piston. In this way, the whole dynamic range of the nonlinear system is tested. Figure 4.14 and Figure 4.15 show the response to different changes in set-point when $N_u = 1$ and $N_u = 2$, respectively. In both cases $\rho = 0.05$ turned out to be a reasonable choice.

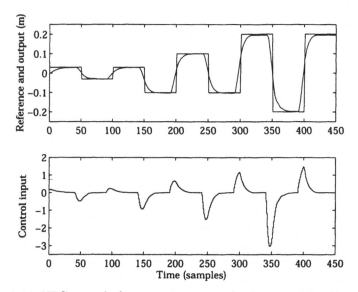

Figure 4.14. NPC control of pneumatic servomechanism. $\rho = 0.05$, $N_u = 1$.

It is seen that the responses are very similar for the two different choices of control horizon and that both control signals and responses are very smooth. The predictive nature of the controller is quite obvious in that the controller clearly anticipates future changes in the set-point. When considering the control signal, it is apparent that one of the nonlinear effects is that the system

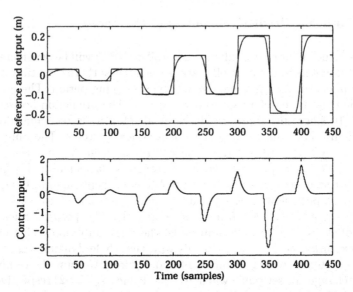

Figure 4.15. NPC control of pneumatic servomechanism. $\rho = 0.05$, $N_u = 2$.

is not symmetric around the position $y = 0$. Nevertheless, the predictive controller is able to produce almost similar responses for all set-point changes.

Unfortunately, the closed-loop system exhibits a small steady-state error for certain set-points despite the fact that the open-loop system contains an integrator. This must be attributed the fact that the network only constitutes an approximate description of the true system. In particular when noticing that the steady-state error is most significant when the magnitude of the set-point is large, this appears to be a plausible explanation. Figure 4.13 shows that large outputs are poorly represented in the training set; thus, it cannot be expected that the network model will behave properly for large outputs.

That the same penalty factor resulted in quite similar responses for three different choices of control horizon indicates that the predictive controller is reasonably insensitive to quite dramatic changes in this design parameter. To examine the effect of changing the penalty factor, a simulation is performed with the penalty reduced by a factor 10, i.e., $\rho = 0.005$. The result is shown in Figure 4.16.

As expected, a smaller penalty factor results in a faster tracking of the reference at the expense of a more active control signal. However, the effect is quite modest considering the large change in penalty.

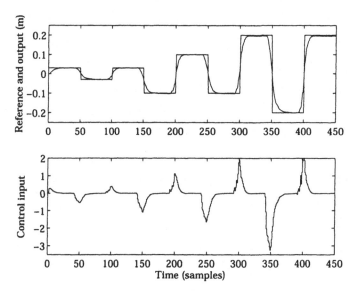

Figure 4.16. The predictive controller is quite insensitive to changes in the penalty factor. In this case $N_u = 2$ and $\rho = 0.005$.

4.3.3 Approximate Predictive Control of the Servo

It is interesting to investigate how the APC will perform on the pneumatic servomechanism since this scheme is much simpler to implement and requires considerably less computations than the NPC. The nonlinearities are reasonably smooth so the main prerequisite for applicability of instantaneous linearization is satisfied. Figure 4.17 shows a simulation performed with the same design parameters as those used in the simulation depicted in Figure 4.15; that is, $N_1 = 1$, $N_2 = 10$, $N_u = 2$, and $\rho = 0.05$.

As expected, the response is similar to the one accomplished with the NPC. However, the steady-state error has been eliminated.

Since the linearized models have been made available as part of the on-line redesign of the controller, one might as well study the variations in the extracted models to gain more insight about the nonlinearities. To get an idea about the magnitude of the nonlinearities, the numerator and denominator coefficients of the extracted models are depicted as functions of time in Figure 4.18.

The variations are very distinct, and they also seem to have a quite systematic correspondence with the change in control input and piston position. To check how the dynamics vary, the location of the poles can be mapped in the complex plane. This is shown in Figure 4.19. As the continuous linear model in (4.3) indicated, the servomechanism has an integrator and a complex pair

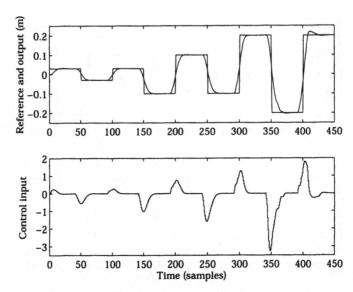

Figure 4.17. The APC applied to the servomechanism.

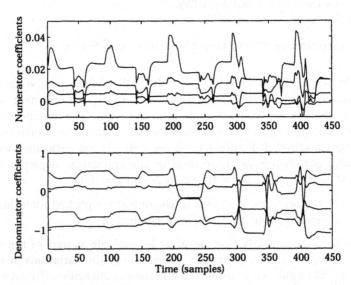

Figure 4.18. Numerator (top) and denominator coefficients (bottom) of the extracted linear models.

of poles. Since an NNARX(4,4,1) is used for modelling the system there is also an additional (real) pole.

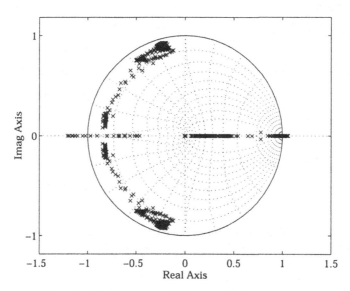

Figure 4.19. The poles of the extracted linear models.

Assume that the linearized models, when considered in continuous time, will comply with the model structure

$$H(s) = \frac{B(s)}{(s + p_1)(s + p_2)(s^2 + 2\zeta\omega s + \omega^2)} \tag{4.4}$$

where $B(s)$ describes the numerator polynomial, where p_1 is always zero (the integration), and where $\omega \simeq 22$ rad/s and $\zeta \simeq 0.034$ for small variations around the equilibrium point for which $y = 0$. When considering the variations of the complex pole pair in Figure 4.19 it is seen that damping and natural frequency vary quite dramatically. By transforming the linearized models to continuous time and separating the poles, the damping factor and natural frequency are easily identified. Figure 4.20 shows how they evolve with time. There appears to be a trend that the damping factor grows with the magnitude of the set-point. The natural frequency, on the other hand, changes primarily when the position is positive.

A somewhat strange behavior is encountered shortly after that 400 samples have been simulated. The complex pole pair reaches the real axis and stays there for a while. It is likely that this is due to inaccuracies in the neural network model. Recall that positions close to the boundary (0.245 meters) are not represented well in the data set; thus, one cannot expect the network

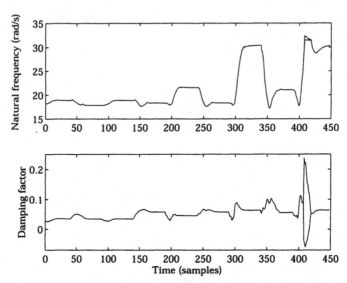

Figure 4.20. Damping factor (top) and natural frequency (bottom). When the poles reach the negative real axis, damping and natural frequency are shown for each pole.

to behave properly here. Such modelling errors can have a fatal impact on the linearized models since these are based on a differentiation of the neural network model.

4.3.4 Section Summary

In this case study the NPC and APC controllers were applied to a model of a pneumatic position servomechanism. The most important conclusions of the study are listed below:

- It is fairly easy to obtain an accurate neural network model when one has access to a large set of training data.

- It can be difficult to collect data in certain regimes when the system is unstable or poorly damped, e.g., near a boundary. It cannot be expected that a neural network model will behave reasonably in regimes that are missing or poorly represented in the training set. Thus, the operating range of the designed closed-loop system should be reduced in order not to violate the validity of the neural network model.

- It was shown that an excellent reference tracking was accomplished by the predictive controllers. It was very easy to tune the controllers to obtain the desired closed-loop behavior.

- When applying the NPC, a steady-state error may occur due to modelling inaccuracies regardless that no external disturbances are acting on the system.

- The APC yields a performance that is similar to that of the NPC, but the controller is much simpler to implement and requires considerably less computations. For the present application the APC was consequently preferred over the NPC.

- The linearized models that come out as a natural bi-product of the APC design are quite useful. Even though the structure of the neural network model is not inspired by the underlying physics of the servomechanism, the instantaneous linearization enables interpretation of the system dynamics.

4.4 Control of Water Level in a Conic Tank

The last example is about controlling the level in a water tank. Examples involving water tanks are used in many textbooks on control system design, e.g., Franklin et al. (1995), and this book will be no exception. To impose a strong nonlinear effect in the system, the water tank is assumed to be shaped like a cone. The system is depicted in Figure 4.21.

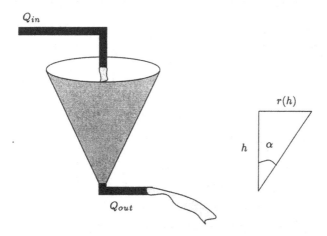

Figure 4.21. The water tank. The control input is the inlet through the pipe above the tank. The output of the system is the water level. There is an (uncontrolled) outlet of water from the bottom of the tank.

The inlet flow, Q_{in}, is the control input to the system, and the output is the water level, h. It is assumed that Q_{in} is controlled directly and that a commanded change of the flow takes effect immediately. The flow is limited to the interval

$$0 \le Q_{in} \le 4 \cdot 10^{-4} \, \text{m}^3/\text{s} \,. \qquad (4.5)$$

Q_{out} is the outlet flow. It is not controlled but is alone due to the hydrostatic pressure. For simplicity it is assumed the flow is turbulent no matter how much water is in the tank; i.e.,

$$Q_{out} = k_{out} \sqrt{h} \,. \qquad (4.6)$$

k_{out} is a constant and has been found to be $k_{out} = 10^{-4} \, \text{m}^{5/2}/\text{s}$.

The height of the water tank is $0.5\,\text{m}$ and the angle $\alpha = 20°$. It is assumed that the water level is available directly. One way of measuring the level in practice is through a measurement of the pressure in the bottom of the tank as this will be proportional to the level.

For the level h the water volume is

$$V(h) = \frac{1}{3}\pi r^2(h)h$$

$$= \left[\frac{1}{3}\pi \tan^2 \alpha\right] h^3 \,. \qquad (4.7)$$

$r(h)$ is here the radius of the tank at the water level; thus, $\tan \alpha = r(h)/h$.

Two expressions for the derivative of the volume can now be written up

$$\frac{dV(h)}{dt} = \begin{cases} Q_{in} - Q_{out} \\ \left[\frac{1}{3}\pi \tan^2 \alpha\right] \dfrac{dh^3}{dt} \end{cases} = \begin{cases} Q_{in} - k_{out}\sqrt{h} \\ \left[\pi \tan^2 \alpha\right] h^2 \dfrac{dh}{dt} \end{cases} \qquad (4.8)$$

leading to the relation

$$\dot{h} = \frac{dh}{dt} = f(h, Q_{in})$$

$$= \frac{1}{\pi \tan^2 \alpha} h^{-2} Q_{in} - \frac{k_{out}}{\pi \tan^2 \alpha} h^{-3/2} \,. \qquad (4.9)$$

4.4.1 Linear Analysis and Control

Before approaching the control problem, it is useful to use a little linear analysis in order to get a feeling for the nonlinearities in the system. Defining, as an equilibrium point, the level $h = h_0$, the corresponding inlet flow will obviously be $Q_{in,0} = k_{out}\sqrt{h_0}$. Linearization of the model about this point gives:

$$a = \frac{\partial f}{\partial h}\bigg|_{h_0, Q_{in,0}} = \frac{-2h_0^{-3}Q_{in,0}}{\pi \tan^2 \alpha} + 3k_{out}h_0^{-5/2}\pi \tan^2 \alpha \qquad (4.10)$$

$$b = \frac{\partial f}{\partial Q_{in}}\bigg|_{h_0, Q_{in,0}} = \frac{h_0^{-2}}{\pi \tan^2 \alpha}, \qquad (4.11)$$

thus leading to

$$\dot{\tilde{h}} \simeq a\tilde{h} + b\tilde{Q}_{in}, \qquad (4.12)$$

where $\tilde{h} \triangleq h - h_0$ and $\tilde{Q}_{in} \triangleq Q_{in} - Q_{in,0}$.

By linearization around two different operating points, one is assured that system is in fact quite nonlinear:

- $(h_0, Q_{in,0}) = (0.03\,\mathrm{m}, 1.73 \cdot 10^{-5}\,\mathrm{m}^3/\mathrm{s})$:

$$a = -0.77, \quad b = 2.7 \cdot 10^3, \quad b/a = -3.5 \cdot 10^3.$$

- $(h_0, Q_{in,0}) = (0.2\,\mathrm{m}, 4.47 \cdot 10^{-5}\,\mathrm{m}^3/\mathrm{s})$:

$$a = -6.7 \cdot 10^{-3}, \quad b = 60, \quad b/a = -8.9 \cdot 10^3.$$

As the level is decreased further below $h = 0.03\,\mathrm{m}$, the dynamic characteristics change extremely rapidly.

Figure 4.22 shows the result when the system controlled by two different linear Generalized Predictive Controllers. The reference takes values between $0.03\,\mathrm{m}$ and $0.4\,\mathrm{m}$. The upper panel shows the performance of a controller designed for the first of the two linearized models while the lower panel shows the performance of a controller designed for the second model. As one would anticipate, the first controller (upper panel) performs well for small levels and poorly for large; for the second controller (lower panel) the opposite result is obtained.

4.4.2 Direct Inverse Control of the Water Level

By following the same approach that was used in Section 4.3 it is straightforward to collect a set of data, train a neural network model, and design an NPC or APC. It is left up to the reader to try this. Instead a direct inverse controller will be attempted this time.

In order to collect a set of training data, the second of the two linear GPCs tried above will be used for conducting the experiment in closed-loop. As usual, care is taken to excite the entire range of operation; that is, there should be both small and big changes in the output and both fast changes

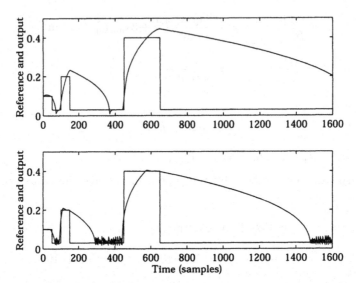

Figure 4.22. GPC control of the water level (only reference and output signals are shown). Upper panel: the controller is designed for the model obtained by linearization about $(h_0, Q_{in,0}) = (0.03\,\text{m}, 1.73 \cdot 10^{-5}\,\text{m}^3/\text{s})$. Lower panel: the controller is designed for the model obtained by linearization about $(h_0, Q_{in,0}) = (0.2\,\text{m}, 4.47 \cdot 10^{-5}\,\text{m}^3/\text{s})$. The following design parameters were used: $N_1 = 1$, $N_2 = 10$, $N_u = 1$, $\rho = 2000$.

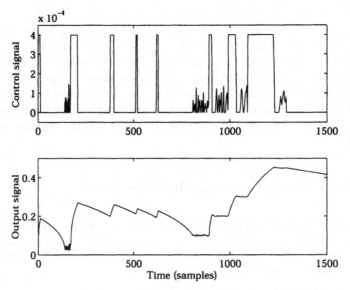

Figure 4.23. The training data. Control signal (upper panel) and output signal (lower panel).

and points where steady-state is reached. A small random signal is added to the control inputs to ensure that the network model will be able to produce reliable high-frequency outputs of small magnitude. The collected training data are shown in Figure 4.23. Before training, the mean is removed from each of the two signals, and the signals are scaled to unit variance.

Although one could make the experiment in open-loop, it is easier to command a trajectory for the output if a controller is available. It is not instrumental that the controller provides great performance. As long as it does not destabilize the system it will generally be useful.

An inverse model is now trained on the acquired data. The model structure is "configured" as shown in Figure 4.24. As the data set is quite large, the

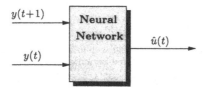

Figure 4.24. Structure of inverse model. When the model has been trained it is used as controller by inserting the reference in place of the future output, $y(t+1)$.

selection of a network architecture is not very critical. A little trial and error testing of few architectures shows that a network with 7 hidden units works well. The network is trained with 500 iterations of the Levenberg-Marquardt algorithm. Only the simple *generalized* training scheme is used as there is no need to proceed with specialized training if an acceptable performance can be achieved without this. The ability of the trained model to predict the true control input is shown in Figure 4.25 for a test set. When looking at the residual it is clear that the error is largest when the control input is either zero or takes its maximum value. This is not surprising as these constraints impose "hard" nonlinearities in the system. A better modelling of the hard nonlinearities generally requires a large network architecture. This may, however, lead to overfitting unless a correspondingly large and informative data set is acquired.

Finally, the inverse model is employed as a controller for the system. The closed-loop response is shown in Figure 4.26.

It is seen that an excellent overall response to step changes is achieved with the inverse controller. For small reference values as well as for large ones. The response is nearly as fast as physically possible and the control signal indicates a "bang-bang" type control. When the reference changes, the maximum or minimum control input is applied until the desired water level is achieved. After that, the control input attains its steady-state value.

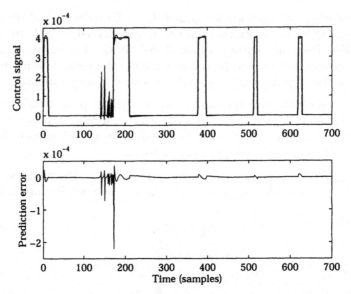

Figure 4.25. Evaluation of the network's ability to predict the true control input. Upper panel: true control signal and the network predictions. Lower panel: prediction errors.

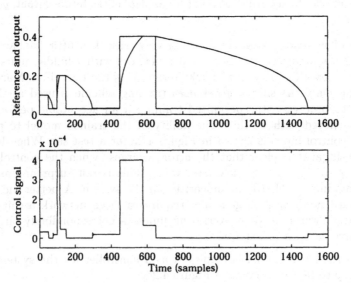

Figure 4.26. Direct inverse control of water level. Upper panel: reference signal and system output. Lower panel: control signal.

4.4.3 Section Summary

In this section it was demonstrated in a simulation study how to design an inverse controller for controlling the level in a water tank. A few remarks on the example are listed below:

- When attempting to design a controller based on a linearized model, one is often put in a dilemma when having to select an operating point about which to linearize. It is difficult to yield a good performance over the entire operating range.

- A poor linear controller can be useful for performing an experiment to acquire data for neural network modelling.

- It is difficult for the neural network to learn hard nonlinearities.

- Direct inverse control can perform well when the inverse model is not unstable or poorly damped. The controller provides a fast response to reference changes.

- Direct inverse controllers are very simple to implement.

References

Ahmed, M. S. and Tasadduq, I. A. (1994). Neural-net controller for nonlinear plants: design approach through linearisation. *IEE Proceedings of Control Theory Applications,* **141**(5).

Åström, K. J., Hagander, P., and Sternby, J. (1984). Zeros of sampled systems. *Automatica,* **20**(1), 31–38.

Åström, K. J. and Wittenmark, B. (1990). *Computer Controlled Systems - Theory and Design.* Prentice Hall, Englewood Cliffs, N.J., 2nd edition.

Åström, K. J. and Wittenmark, B. (1995). *Adaptive Control.* Addison-Wesley, Reading, MA, 2nd edition.

Barron, A. R. (1993). Universal approximation bounds for superpositions of a sigmoid function. *IEEE Transactions on Information Theory,* **39**, 930–945.

Bierman, G. J. (1977). *Factorization Methods for Discrete Sequential Estimation.* Academic Press, London, UK.

Billings, S. A., Jamaluddin, H. B., and Chen, S. (1992). Properties of neural networks with applications to modelling non-linear dynamical systems. *International Journal of Control,* **55**(1), 193–224.

Billings, S. A. and Voon, W. S. F. (1986). Correlation based model validity tests for non-linear models. *International Journal of Control,* **44**(1), 235–244.

Billings, S. A. and Zhu, Q. M. (1994). Nonlinear model validation using correlation tests. *International Journal of Control,* **60**(6), 1107–1120.

Bittani, S. and Piroddi, L. (1993). A neural network approach to generalized minimum variance control of nonlinear systems. In Nieuwenhuis, J., Praagman, C., and Trentelman, H., editors, *Proc. European Control Conference, Groningen,* pp. 466–470, Groningen, Netherland.

Chen, F. and Khalil, H. K. (1991). Adaptive control of nonlinear systems using neural networks - a dead-zone approach. In *Proceedings of the American Control Conference, Boston, MA,* pp. 667–672.

Clarke, D. and Gawthrop, P. (1975). Self-tuning controller. *IEE Proceedings D, Control Theory and Application,* **9**(122), 929–934.

Clarke, D. W., Mothadi, C., and Tuffs, P. S. (1987a). Generalized predictive control - Part I. The basic algorithm. *Automatica,* **23**(2), 137–148.

Clarke, D. W., Mothadi, C., and Tuffs, P. S. (1987b). Generalized predictive control - Part II. Extensions and interpretations. *Automatica*, **23**(2), 149–160.

Craig, J. J. (1989). *Introduction to Robotics - Mechanics and Control.* Addison-Wesley, Reading, MA, 2nd edition.

Cybenko, G. (1989). Approximation by superpositions of a sigmoidal function. *Mathematics of Control, Signals, and Systems*, **2**(4), 303–314.

Demuth, H. and Beale, M. (1998). *Neural Network Toolbox: User's Guide, Version 3.0.* The MathWorks, Inc., Natick, MA.

Dennis, J. E. and More, J. J. (1977). Quasi-Newton methods, motivation and theory. *SIAM Review*, **19**(1), 46–89.

Dennis, J. E. and Schnabel, R. B. (1983). *Numerical Methods for Unconstrained Optimization and Nonlinear Equations.* Prentice-Hall, Englewood Cliffs, NJ.

Eaton, J. W., Rawlings, J. B., and Ungar, L. H. (1994). Stability of neural net based model predictive control. In *Proc. of the American Control Conference, Baltimore, Maryland, USA*, pp. 2481–2485.

Efron, B. and Tibshirani, R. (1993). *An Introduction to the Bootstrap.* Chapman & Hall, New York.

Fletcher, R. (1987). *Practical Methods of Optimization.* Wiley & Sons, Chichester, 2nd edition.

Fog, T. L., Larsen, J., and Hansen, L. K. (1995). Training and evaluation of neural networks for multi-variate time seriesd processing. In *Proc. IEEE International Conference on Neural Networks, Perth, Australia*, pp. 1194–1199.

Franklin, G. F., Powell, J. D., and Emami-Naeini, E. A. (1995). *Feedback Control of Dynamic Systems.* Addison-Wesley, Reading, MA, 3rd edition.

Franklin, G. F., Powell, J. D., and Workman, M. L. (1998). *Digital Control of Dynamic Systems.* Addison-Wesley, Reading, MA, 3rd edition.

Gawthrop, P. (1977). Some interpretations of the self-tuning controller. *IEE Proceedings D, Control Theory and Application*, **1**(124), 889–894.

Geman, S., Bienenstock, E., and Doursat, R. (1992). Neural networks and the bias/variance dilemma. *Neural Computation*, **4**(1), 1–58.

Gevers, M. (1993). Towards a joint design of identification and control. In Trentleman, H. L. and Willems, J. C., editors, *Essays on Control: Perspectives in the Theory and its Applications*, volume 14 of *Progress in Systems and Control Theory*, chapter 5, pp. 111–151. Birkhäuser, Boston, MA.

Gorodkin, J., Hansen, L., Krogh, A., Svarer, C., and Winther, O. (1993). A quantitative study of pruning by optimal brain damage. *International Journal of Neural Systems*, **4**(2), 159–169.

Grewal, M. S. and Andrews, A. P. (1993). *Kalman Filtering: Theory and Practice.* Prentice Hall Information and System Science series. Prentice-Hall, Englewood Cliffs, NJ.

Gunnarsson, S. and Krus, P. (1997). fluid power control of a flexible mechanical structure. Technical Report LiTH-ISY-R-1961, Dept. of EE. Linköping University, S-581 83 Linköping, Sweden.

Haber, R. (1985). Nonlinearity tests for dynamic processes. In Barker, H. A. and Young, P. C., editors, *Proc. IFAC Symp. on Identification and System Parameter Estimation, York, UK*, pp. 409–414.

Hagan, M. T. and Menhaj, M. B. (1994). Training feedforward networks with the Marquardt algorithm. *IEEE Transactions on Neural Networks*, **5**(6), 989–993.

Haley, P., Soloway, D., and Gold, B. (1999). Real-time adaptive control using neural generalized control. In *Proc. 1999 American Control Conference, San Diego, California*, pp. 4278–4282.

Hansen, L. K. and Larsen, J. (1996). Linear unlearning for cross-validation. *Advances in Computational Mathematics*, **5**, 269–280.

Hansen, L. K. and Pedersen, M. W. (1994). Controlled growth of cascade correlation nets. In Marinaro, M. and Morasso, P. G., editors, *Proc. ICANN'94, Sorrento, Italy*, pp. 797–800.

Hansen, L. K., Rasmussen, C. E., Svarer, C., and Larsen, J. (1994). Adaptive regularization. In Vlontzos, J., Hwang, J.-N., and Wilson, E., editors, *Proc. IEEE Workshop on Neural Networks for Signal Processing IV, Piscataway, New Jersey*, pp. 78–87.

Hansen, L. K. and Salamon, P. (1990). Neural network ensembles. *IEEE Transactions on Pattern Analysis and Machine Intelligence*, **12**, 993–1001.

Hassibi, B. and Stork, D. G. (1993). Second derivatives for network pruning: optimal brain surgeon. In Hanson, S. J., Cowan, J. D., and Giles, C. L., editors, *Advances in Neural Information Processing Systems 5. Proceedings of the 1992 Conference*, pp. 164–171, San Mateo, CA. Morgan Kaufmann.

Haykin, S. (1998). *Neural networks: A Comprehensive Foundation*. Prentice Hall, 2nd edition.

He, X. and Asada, H. (1993). A new method for identifying orders of input-output models for nonlinear dynamical systems. In *Proc. of the American Control Conference, San Francisco, California*, pp. 2520–2523.

Hertz, J., Krogh, A., and Palmer, R. G. (1991). *An Introduction to the Theory of Neural Computation*. Lecture Notes, Volume I. Addison-Wesley, Redwood City, CA.

Heskes, T. (1998). Bias/variance decompositions for likelihood-based estimators. *Neural Computation*, **10**(6), 1425–1433.

Hunt, K. J. and Sbarbaro-Hofer, D. (1991). Neural networks for nonlinear internal model control. *IEE Proc. D*, **138**(5), 431–438.

Isermann, R., Lachmann, K., and Matko, D. (1992). *Adaptive Control Systems*. Systems and Control Engineering. Prentice-Hall, New York, NY.

Isidori, A. (1995). *Nonlinear Control Systems*. Springer-Verlag, London, UK, 3rd edition.

Jagannathan, S. and Lewis, F. L. (1996). Multilayer discrete-time neural-net controller with guaranteed performance. *IEEE Transactions on Neural Networks*, **7**(1), 107–130.

Jin, L., Nikiforuk, P. N., and Gupta, M. M. (1992). Adaptive tracking of SISO nonlinear systems using neural networks. In *Proc. of the American Control Conference, Chicago, Illinois*, pp. 56–60.

Juditsky, A., Hjalmarson, H., Benveniste, A., Delyon, B., Ljung, L., Sjöberg, J., and Zang, Q. (1995). Nonlinear black-box models in system identification: Mathematical foundations. *Automatica*, **31**(12), 1725–1750.

Khalil, H. K. (1996). *Nonlinear Systems*. Prentice Hall.

Kučera, V., Ječek, J., and Krupička, M. (1991). Numerical analysis of Diophane equations. In Warwick, K., Kárný, M., and Halousková, A., editors, *Advanced Methods in Adaptive Control for Industrial Applications*, number 158 in Lecture Notes in Control and Information Sciences, pp. 128–136. Springer-Verlag, Berlin.

Kwakernaak, H. and Sivan, H. (1972). *Linear Optimal Control Systems*. John Wiley and Sons, New York, NY.

Larsen, J. (1993). *Design of Neural Network Filters*. PhD thesis, Electronics Institute, Technical University of Denmark, Kgs. Lyngby, Denmark.

Larsen, J. and Hansen, L. (1994). Generalization performance of regularized neural network models. In Vlontzos, J., Whang, J.-N., and Wilson, E., editors, *Proceedings of the 4th IEEE Workshop on Neural Networks for Signal Processing*, pp. 42–51, Piscataway, NJ.

Larsen, J., Svarer, C., Andersen, L. N., and Hansen, L. K. (1998). Adaptive regularization in neural network modeling. In Orr, G. B. and Moeller, K. R., editors, *Neural Networks: Tricks of the Trade*, volume Lecture Notes in Computer Science 1524, pp. 113–132, germany. Springer-Verlag.

Le Cun, Y., Denker, J., and Solla, S. (1990). Optimal brain damage. In Touretzky, D. S., editor, *Neural Information Processing Systems*, volume 2, pp. 598–605, San Mateo. (Denver 1989), Morgan Kaufmann.

Le Cun, Y., Kanter, I., and Solla, S. A. (1991). Eigenvalues of covariance matrices: application to neural-network learning. *Physical Review Letters*, **66**(18), 2396–2399.

Levenberg, K. (1944). A method for solution of certain nonlinear problems in least squares. *Quart. Appl. Mathematics*, **2**, 164–168.

Lightbody, G. and Irwin, G. (1995). A novel neural internal model control structure. In *Proc. the American Control Conference, Seattle, Washington*, pp. 350–354.

Ljung, L. (1999). *System Identification - Theory for the User*. Prentice Hall, Upper Saddle River, N.J., 2nd edition.

Ljung, L. and Sjöberg, J. (1992). A system identification perspective on neural nets. Technical Report LiTH-ISY-I-1373, Division of Automatic Control Systems, Linköping, University, Linköping, Sweden.

Ljung, L. and Söderström, T. (1983). *Theory and Practice of Recursive Identification*. MIT Press, Cambridge, MA.

MacKay, D. (1992a). Bayesian interpolation. *Neural Computation*, **4**(3), 415–447.

MacKay, D. (1992b). A practical bayesian framework for backpropagation networks. *Neural Computation*, **4**(3), 448–472.

Madsen, K. (1984). *Lecture Notes on Optimization Without Contraints*. Technical University of Denmark, Department of Mathematical Modelling, Lyngby, Denmark. In Danish.

Madsen, P. P. (1995). Neural network for optimization of existing control systems. In *Proc. 1995 IEEE International Conference on Neural Networks, Perth, Australia*, pp. 1496–1501.

Marquardt, D. (1963). An algorithm for least-squares estimation of nonlinear parameters. *SIAM Journal Appl. Mathematics*, **11**(2), 164–168.

Mayne, D. Q. and Michalaska, H. (1990). Receding horizon control of nonlinear systems. *IEEE Transactions on Automatic Control*, **35**(7), 814–824.

Middleton, R. and Goodwin, G. (1990). *Digital Control and Estimation: a Unified Approach*. Prentice Hall, Englewood Cliffs, NJ.

Mohtadi, C. (1988). Numerical algorithms in self-tuning control. In Warwick, K., editor, *Implementation of Self-tuning Controllers*, pp. 67–95. Peter Peregrinus, London.

Moody, J. (1991). Note on generalization, regularization and architecture selection in nonlinear learning systems. In B.H. Juang, S.Y. Kung, C. K., editor, *Proceedings of the First IEEE Workshop on Neural Networks for Signal Processing*, p. 1:10. IEEE, Piscataway, NJ.

Moody, J. (1994). Prediction risk and architecture selection for neural networks. In Cherkassky, V., Friedman, J. H., and Wechsler, H., editors, *From Statistics to Neural Networks: Theory and Pattern Recognition Applications*, volume Series F136. Germany: Springer-Verlag.

Morari, M. and Zafiriou, E. (1989). *Robust Process Control*. Prentice-Hall, Englewood Cliffs, NJ.

Mørch, N., Hansen, L. K., Strother, S. C., Svarer, C., Rottenberg, D. A., Lautrup, B., Savoy, R., and Paulson, O. B. (1997). Nonlinear vs. linear models in functional neuroimaging: learning curves and generalization crossover. In Duncan, J. and Gindi, G., editors, *In Information Processing in Medical Imaging*, volume 7, pp. 259–270. Springer-Verlag, VT.

Moré, J. J. (1983). Recent developments in algorithms and software for trust-region methods. In *Mathematical Programming, the State of the Art: Bonn 1982*. Springer-Verlag, Berlin, Germany.

Narendra, K. S. and Parthasarathy, K. (1992). Identification and control of dynamic systems using neural networks. *IEEE Trans. on Neural Networks*, **1**(1), 4–27.

Nijmeijer, H. and van der Schaft, A. J. (1990). *Nonlinear Dynamical Control Systems*. Springer-Verlag, New York, NY.

Nørgaard, M. (1996a). Neural network based control system design toolkit. Tech. report 96-E-830, Department of Automation, DTU, Lyngby, Denmark.

Nørgaard, M. (1996b). *System Identification and Control with Neural Networks*. PhD thesis, Department of Automation, Technical University of Denmark, Kgs. Lyngby, Denmark.

Nørgaard, M. (1997). Neural network based system identification toolbox. Tech. report 97-E-851, Department of Automation, DTU, Lyngby, Denmark.

Parkum, J., Poulsen, N. K., and Holst, J. (1992). Recursive forgetting algorithms. *International Journal of Control*, **55**(1), 109–128.

Pedersen, M. W. and Hansen, L. K. (1995). Recurrent networks: second order properties and pruning. In Tesauro, G., Touretzky, D., and Leen, T., editors, *Advances in Neural Information Processing Systems*, volume 7, pp. 673–680. The MIT Press.

Press, W. H., Flannary, B. P., Teukolsky, S. A., and Vetterling, W. T. (1988). *Numerical Recipes in C*. Cambridge University Press, Cambridge, UK.

Priestley, M. B. (1988). *Non-linear and Non-stationary Time Series Analysis*. Academic Press, New York, USA.

Pröll, T. and Karim, M. N. (1994). Real-time design of an adaptive nonlinear predictive controller. *International Journal of Control*, **59**(3), 863–889.

Psaltis, D., Sideris, A., and Yamamura, A. A. (1988). A multilayer neural network controller. *IEEE Control Systems Magazine*, **8**(2), 17–21.

Rumelhart, D. E. and McClelland, J. L. (1986). *Parallel Distributed Processing: Explorations in the Microstructure of Cognition: Foundations*, volume 1. MIT Press, Cambridge, MA.

Salgado, M., Goodwin, G., and Middleton, R. (1988). Modified least squares algorithm incorporating exponential forgetting and resetting. *Int. Journal of Control*, **47**(2), 477–491.

Salle, J. L. and Lefschetz, S. (1961). *Stability by Lyaponov's Direct Method*. Mathematics in Science and Engineering. Academic Press.

Sanner, R. M. and Slotine, J.-J. E. (1992). Gaussian networks for direct adaptive control. *IEEE Transactions on Neural Networks*, **3**(6), 837–863.

Sarle, W. S. (1994). Neural networks and statistical models. In *Proc. 19th Annual SAS User Group Int. Conference, Cary,NC*, pp. 1538–1550. SAS Institute.

Seber, G. A. F. and Wild, C. J. (1989). *Nonlinear Regression*. John Wiley & Sons, New York, NY.

Sjöberg, J. and Ljung, L. (1995). Overtraining, regularization, and searching for minimum in neural networks. *Int. Journal of Control*, **62**(6), 1391–1408.

Sjöberg, J., Zhang, Q., Ljung, L., Benveniste, A., B. Delyon, P.-Y. Glorennec, H. H., and Juditsky, A. (1995). Nonlinear black-box modeling in system identification: a unified overview. *Automatica*, **31**(12), 1691–1724.

Slotine, J.-J. E. and Li, W. (1991). *Applied Nonlinear Control.* Prentice-Hall, Englewood Cliffs, N.J.

Söderström, T. and Stoica, P. (1989). *System Identification.* Prentice-Hall, London, UK.

Soeterboek, R. (1992). *Predictive Control - A Unified Approach.* Prentice Hall, New York, NY.

Soloway, D. and Haley, P. J. (1996). Neural/generalized predicitve control, a Newton-Raphson implementaion. In *Proceedings of the 11th IEEE int. Symposium on Intelligent Control,* pp. 277–282.

Sørensen, O. (1993). Neural networks performing system identification for control applications. In *3rd Int. Conference on Artificial Neural Networks, Brighton, UK,* pp. 172–176. IEE.

Sørensen, O. (1994). *Neural Networks in Control Applications.* PhD thesis, Aalborg University, Department of Control Engineering, Aalborg, Denmark.

Sørensen, O. (1996). Non-linear pole-placement control with a neural network. *European Journal of Control,* 2(1).

Sørensen, P. H., Nørgaard, M., Larsen, J., and Hansen, L. (1996). Cross-validation with LULOO. In Amari, S., Xu, L., Chan, L.-W., King, I., and Leung, K.-S., editors, *Proc. of the 1996 Int. Conference on Neural Information Processing, Hong-Kong,* volume 2, pp. 1305–1310, Singapore. Springer-Verlag.

Sørensen, P. H., Sinha, P. K., and Al-Mutib, K. (1994). Identification of a pneumatic servo mechanism using neural networks. In *Proc. International Conference on Machine Automation,* volume 2, pp. 499–512, Tampere, Finland.

Svarer, C., Hansen, L. K., and Larsen, J. (1993). On design and evaluation of tapped-delay neural network architectures. In *1993 IEEE Int. Conf. on Neural Networks, San Francisco, CA, USA,* pp. 45–51.

Tong, H. (1990). *Nonlinear Time Series, a Dynamical Systems Approach.* Oxford, UK, Clarendon Press.

Toussaint, G. T. (1974). Bibliography on estimation of misclassification. *IEEE Transactions on Information Theory,* 20(4), 472–479.

Tzirkel-Hancock, E. and Fallside, F. (1992). Stable control of nonlinear systems using neural networks. *International Journal of Robust and Nonlinear Control,* 2(1), 63–86.

Wahba, G. (1990). *Spline Models for Observational Data.* SIAM, Philadelphia, PA.

Weigend, A. S., Huberman, B. A., and Rumelhart, D. E. (1990). Predicting the future: a connectionist approach. *Int. J. of Neural Systems,* 3(1), 193–209.

Xu, J.-X., Donne, J., and Özgüner, U. (1991). Synthesis of feedback linearization and variable structure control with neural net compensation. In *Proc.*

1991 IEEE Int. Symposium on Intelligent Control, Arlington, VA, USA, pp. 184–189.

Yesildirek, A. and Lewis, F. L. (1994). Feedback linearization using neural networks. In *Proc. 1994 IEEE Int. Conference on Neural Networks, Orlando, FL, USA*, pp. 2539–2544.

Ziegler, J. G. and Nichols, N. B. (1942). Optimum settings for automatic controllers. *Trans. ASME*, **64**, 759–768.

Zurada, J. M. (1992). *Artificial Neural Systems*. West Publishing Company, St. Paul, MN.

Index